Collin Mulliner

Mobile Phone Security

Collin Mulliner

Mobile Phone Security

The Impact of the Cellular Modem

Südwestdeutscher Verlag für Hochschulschriften

Impressum/Imprint (nur für Deutschland/only for Germany)
Bibliografische Information der Deutschen Nationalbibliothek: Die Deutsche Nationalbibliothek verzeichnet diese Publikation in der Deutschen Nationalbibliografie; detaillierte bibliografische Daten sind im Internet über http://dnb.d-nb.de abrufbar.

Alle in diesem Buch genannten Marken und Produktnamen unterliegen warenzeichen-, marken- oder patentrechtlichem Schutz bzw. sind Warenzeichen oder eingetragene Warenzeichen der jeweiligen Inhaber. Die Wiedergabe von Marken, Produktnamen, Gebrauchsnamen, Handelsnamen, Warenbezeichnungen u.s.w. in diesem Werk berechtigt auch ohne besondere Kennzeichnung nicht zu der Annahme, dass solche Namen im Sinne der Warenzeichen- und Markenschutzgesetzgebung als frei zu betrachten wären und daher von jedermann benutzt werden dürften.

Coverbild: www.ingimage.com

Verlag: Südwestdeutscher Verlag für Hochschulschriften GmbH & Co. KG
Heinrich-Böcking-Str. 6-8, 66121 Saarbrücken, Deutschland
Telefon +49 681 37 20 271-1, Telefax +49 681 37 20 271-0
Email: info@svh-verlag.de

Approved by: Berlin, TU, Diss., 2011

Herstellung in Deutschland (siehe letzte Seite)
ISBN: 978-3-8381-3289-1

Imprint (only for USA, GB)
Bibliographic information published by the Deutsche Nationalbibliothek: The Deutsche Nationalbibliothek lists this publication in the Deutsche Nationalbibliografie; detailed bibliographic data are available in the Internet at http://dnb.d-nb.de.

Any brand names and product names mentioned in this book are subject to trademark, brand or patent protection and are trademarks or registered trademarks of their respective holders. The use of brand names, product names, common names, trade names, product descriptions etc. even without a particular marking in this works is in no way to be construed to mean that such names may be regarded as unrestricted in respect of trademark and brand protection legislation and could thus be used by anyone.

Cover image: www.ingimage.com

Publisher: Südwestdeutscher Verlag für Hochschulschriften GmbH & Co. KG
Heinrich-Böcking-Str. 6-8, 66121 Saarbrücken, Germany
Phone +49 681 37 20 271-1, Fax +49 681 37 20 271-0
Email: info@svh-verlag.de

Printed in the U.S.A.
Printed in the U.K. by (see last page)
ISBN: 978-3-8381-3289-1

Copyright © 2012 by the author and Südwestdeutscher Verlag für Hochschulschriften GmbH & Co. KG and licensors
All rights reserved. Saarbrücken 2012

Contents

1	**Introduction**		**7**
	1.1	Contributions	9
		1.1.1 Vulnerability Analysis for Mobile Handsets	9
		1.1.2 Cellular Malware Communication Capabilities	10
		1.1.3 Mitigation of Cellular Signaling Attacks on Smartphones	10
	1.2	Structure of the Work	11
2	**Background**		**13**
	2.1	Cellular Network Architecture	13
	2.2	Cellular Signaling	15
	2.3	The Short Message Service	16
	2.4	Smartphone Architecture	16
3	**Security Analysis of SMS Clients on Smartphones**		**19**
	3.1	Introduction	19
	3.2	Related Work	21
	3.3	The SMS_DELIVER Format	22
	3.4	Mobile Phone Side SMS Delivery	23
		3.4.1 The Telephony Stack	23
		3.4.2 SMS Delivery	24
		3.4.3 The Stacks of our Test Devices	24
	3.5	SMS Injection	25
		3.5.1 The Injection Framework	26
	3.6	Fuzzing SMS Implementations	28
		3.6.1 Fuzzing Test Cases	28
		3.6.2 Delivery	30
		3.6.3 Monitoring	30

Contents

      3.6.4   Fuzzing Results 30
  3.7 Conclusions 31

4 Security Analysis of SMS Implementations on Feature Phones   33
  4.1 Introduction 33
  4.2 Related Work 35
  4.3 Target Selection 37
  4.4 Security Analysis of Feature Phones 40
      4.4.1   Network Setup 40
      4.4.2   Sending SMS Messages 42
      4.4.3   Monitoring for Crashes 43
      4.4.4   Additional Monitoring Techniques 46
      4.4.5   SMS_SUBMIT Encoding 47
      4.4.6   Fuzzing Test-cases 49
      4.4.7   Fuzzing Trial 50
      4.4.8   Results 51
      4.4.9   Validation and Extended Testing 53
      4.4.10 Bug Characterization 54
  4.5 Implementing the Attack 54
      4.5.1   Building a Hit-List 54
      4.5.2   Sending SMS Messages 55
      4.5.3   Reducing the Number of Messages 56
      4.5.4   Network Assisted Attack Amplification 57
      4.5.5   Attack Scenarios and Impact 59
  4.6 Countermeasures 60
      4.6.1   Detection 60
      4.6.2   SMS Filtering 61
      4.6.3   Preventing Network Amplification 62
  4.7 Conclusions 62

5 Cellular Malware Communication Capabilities   65
  5.1 Introduction 65
  5.2 Howto hijack many thousand iPhones 67
  5.3 Cellular Challenges 68
      5.3.1   Absence of Public IP Addresses 68

	5.3.2	Platform Diversity	68
	5.3.3	Constant Change of Connectivity	69
	5.3.4	Communication Costs	69
5.4	C&C for Mobile Botnets	70
	5.4.1	Securing the C&C Communication	71
	5.4.2	Peer-to-Peer C&C	71
	5.4.3	SMS C&C	72
5.5	Communication Strategies	77
	5.5.1	IP-based Communication	77
	5.5.2	SMS-based Communication	78
	5.5.3	Data Delivery	78
5.6	Bot Implementation	79
	5.6.1	Commands	79
	5.6.2	Kademlia P2P Client	80
	5.6.3	SMS Client	80
	5.6.4	Evaluation	81
5.7	Conclusions	83

6 Improving the Security of the Cellular Modem Interface   85

6.1	Introduction	85
6.2	Threats	87
	6.2.1	Hijacked Phones and Mobile Botnets	87
	6.2.2	PDP Context Change	88
	6.2.3	Premium Rate SMS Trojans	88
	6.2.4	Rooted Phones	89
6.3	Design	89
	6.3.1	Micro Kernel as Secure Foundation	91
	6.3.2	Virtualized Android	92
	6.3.3	Virtual Modem	93
6.4	Implementation	94
	6.4.1	Hardware	94
	6.4.2	L4Android	94
	6.4.3	System Setup	95
	6.4.4	Modifications to the Android RIL	97
6.5	The AT Command Filter	98

		6.5.1	AT Command Characterization	98
		6.5.2	PDP Context Setup on the STE Baseband	99
		6.5.3	Special Problems	100
		6.5.4	Filtering AT Commands	101
		6.5.5	SMS Filter	102
		6.5.6	Blocking Commands	102
		6.5.7	Profiling benign AT Command Usage	103
	6.6	Evaluation		105
		6.6.1	Our GSM Test Network	105
		6.6.2	Limiting the Call-forwarding Attack	105
		6.6.3	Limiting PDP Context Changes	106
		6.6.4	SMS Trojan	107
		6.6.5	SMS Controlled Botnets	108
	6.7	Related Work		108
	6.8	Conclusions		109
		6.8.1	Future Improvements	110

7 Conclusions   113

Acknowledgements   117

List of Figures   119

List of Tables   121

Bibliography   123

Papers

Published Papers

Parts of this work are based on the following peer-reviewed papers that have already been published.

Conferences

C. MULLINER, S. LIEBERGELD, M. LANGE, AND J.-P. SEIFERT Taming Mr Hayes: Mitigating Signaling Based Attacks on Smartphones. In *Proceedings of the IEEE/IFIP 41st International Conference on Dependable Systems Networks (DSN)* (2012).

C. MULLINER, N. GOLDE, AND J.-P. SEIFERT SMS of Death: from analyzing to attacking mobile phones on a large scale. In *Proceedings of the 20th USENIX Security Symposium San Francisco (Security)* (2011).

C. MULLINER AND J.-P. SEIFERT Rise of the iBots: 0wning a telco network. In *Proceedings of the 5th IEEE International Conference on Malicious and Unwanted Software (Malware)* (2010).

C. MULLINER AND C. MILLER Injecting SMS Messages into Smart Phones for Security Analysis. In *Proceedings of the 3rd USENIX Workshop on Offensive Technologies (WOOT)* (2009).

1 Introduction

Cellular communication and mobile phones are deeply anchored in our daily lives as we rely on them every day, equally, for work and leisure. With this heavy reliance, security and reliability become a necessity. Unfortunately, mobile phones and cellular communication in general are complex and securing them is a challenging task. The task becomes even more challenging as sophisticated attacks (with clear financial goals) are becoming common place. Nowadays, mobile malware [53] brings the necessary capabilities to gain full control over the infected phones and thus has unlimited access to stored data and the cellular network.

Our society's strong dependence on reliable and secure mobile communication has sparked a lot of research effort in order to secure mobile handsets and cellular networks, alike. The security of cellular networks has been studied and improvements have been suggested [26, 84–86]. In this work, we investigate the security of mobile handsets.

Mobile handsets have many interesting properties when it comes to the security of the whole cellular infrastructure. Mobile handsets come in many different flavors. From very simple phones that can only make voice calls to high end smartphones that have more storage and computing power than a laptop computer from a few years ago. Especially the capabilities of high end smartphones can pose a real threat to the user and the cellular network. An example is the rich API set that exposes most of the hardware capabilities to the hosted applications. To counter threats arising from mobile malware and hijacked handsets a plethora of research has been carried out to understand and demonstrate the security issues of cellular handsets [37, 45, 57, 58, 62, 63, 81]. Likewise, effort has gone into improving the security of mobile handsets [25, 64, 94].

However, so far the majority of work has been conducted on the operating system and application side of mobile phones whereas the cellular modem was neglected. The cellular modem is one of the essential parts of cellular phones since it provides the interface to the cellular network. The negligence most likely roots from the closed nature of cellular modems where manufacturers regard every detail as a trade secret.

1 Introduction

Hence, only few studies [91] have looked into cellular modem security. The Federal Communication Commission (FCC) also raised some security concerns related to the cellular modem [30] but did not further enforce any regulations. Because of these shortcomings we investigate the security at the boundary of the cellular modem and the mobile operating system.

This work investigates the impact of the cellular modem on the security of mobile phones. The goal is to understand the security issues and shortcomings of the modem interface and to propose systematic improvements to it. We formulate the following hypothesis:

Securing the cellular modem interface improves the security of mobile handsets and the cellular infrastructure.

We further claim that:

(i) The cellular modem plays a significant role in securing mobile handsets.

(ii) Access and usage of the cellular modem is so far only coarse-grained.

(iii) Unhindered access to the cellular modem presents a security threat to both, the network and the individual user.

We begin our investigation by analyzing the cellular modem interface and how it interacts with the mobile operating system. We interpose between the cellular modem interface and telephony software stack on current smartphone platforms. Using this capability, we develop a method for vulnerability analysis for Short Message Service (SMS) implementations. Our analysis shows that the telephony stack of many modern smartphone platforms contain security critical software bugs.

Following up on the first part, we continue the vulnerability analysis of SMS implementations on feature phones. Feature phones are mobile phones that have additional features besides voice calls and text messaging such as a web browser. Feature phones are of particular interest because of their large field deployment (smartphones only account for about 16% of all mobile phones [11]) and the fact that feature phones only consists of a single CPU. The single CPU implements both the cellular modem as well as the mobile phone operating system and applications (smartphones normally have a dedicated CPU for each task). The single CPU architecture bears the potential of security and reliability issues due to unwanted side effects between the software module that implements the cellular modem and the module that implements the operating system. We conceive an analysis method and tool to investigate feature phones. We determine and show that SMS implementations of all major feature phone platforms contain security relevant issues.

Now that we understand the interaction between the cellular modem and the operating system we investigate malicious modem access. Specifically, we address the question about the capabilities of malicious software that has direct access to the cellular modem. Therefore, we create a Proof-of-Concept mobile botnet where the bots utilize the access to the modem to silently communicate. Here we demonstrate that such a botnet is practically feasible and discuss details of such communication.

The final part of this work concentrates on possible countermeasures to prevent malicious access to the cellular modem. We design a protection system to control access to the cellular modem. Our system prevents unlimited access to the modem by malware that is executed on the mobile operating system. The evaluation of our protection system shows that it can mitigate different kinds of malicious access such as SMS-based botnet communication and cellular signaling attacks.

1.1 Contributions

The work provides an insight on the impact of the cellular modem interface to the security of mobile phones. The work makes several contributions to the area of cellular security and its main contributions are the following.

1.1.1 Vulnerability Analysis for Mobile Handsets

Vulnerability analysis of mobile phones is a challenging task especially if the targets are the software modules that interact with the cellular network. Analyzing and testing these components for security defects is challenging because of the cellular network infrastructure that is required in order to stimulate the target software module. Cellular networks are large, unpredictable, and slow and thus unsuited to produce reliable and reproducible test results.

We solve these challenging tasks and design vulnerability analysis methods that besides other properties cut out the complicated cellular network infrastructure. On smartphones we slightly modify the cellular software stack to interpose between the cellular

1 Introduction

modem and the application processor. This allows for emulating rogue behavior of the cellular network. The stack modification relies on an ubiquitous design that is common to most smartphones.

For mobile phones that do not allow software modifications, we develop an analysis method that is based on a private GSM base station. The base station is operated by our custom (open source) software. This setup not only allows sending messages to phones, to trigger software faults, but also is able to collect feedback from the phones. Based on that feedback we create a method that uses the private base station as a debugging tool for mobile phones.

1.1.2 Cellular Malware Communication Capabilities

Mobile malware is on the rise. To better understand it and to provide insights into the capabilities of malware residing on cellular phones, we create a Proof-of-Concept cellular botnet. We mainly focus on the command and control communication since this is the primary toehold for fighting botnets.

We design, implement, and evaluate several communication methods for cellular botnets. We emphasize stealthy SMS-based communication through direct access to the cellular modem interface by the bots. Our work shows that stealthy SMS-based communication is feasible and that mobile network operators have the best vantage point for detecting and fighting cellular bots that purely leverage the cellular infrastructure for communication.

1.1.3 Mitigation of Cellular Signaling Attacks on Smartphones

Cellular signaling attacks carried out by compromised smartphones are a serious security threat to cellular networks. The attacks are based on flooding the network with specific messages that create a high load on various network components and thus cause specific parts of cellular networks to fail. Attacks like these have been demonstrated in the past. Such attacks are carried out by hijacked smartphones where the hijacker or malware has unlimited access to the cellular modem interface.

We propose a protection system for smartphones that prevents malware from abusing the cellular modem interface for carrying out signaling attacks. Evaluation of our protection system shows that it can limit the rate of harmful signaling messages that a phone can send towards the network and thus protect cellular infrastructure and its users.

1.2 Structure of the Work

The rest of the work is structured as follows: Chapter 2 provides necessary background information on cellular communication systems. In Chapters 3 and 4 we present two studies on vulnerability analysis methods of feature phones and smartphones. The studies themselves focus on SMS implementations. Chapter 5 presents a study on the communication capabilities of cellular malware under the premise that the malware has direct access to cellular modem interface. This chapter focuses on implementing a cellular botnet that mainly communicates via SMS messaging. Chapter 6 introduces a security architecture for smartphones to mitigate malicious access to the cellular modem from the application processor. The architecture presents a method to protect cellular networks from compromised smartphones. In Chapter 7, we summarize the presented work, draw our conclusions, and briefly discuss potential future work.

1 Introduction

2 Background

In this chapter we provide the necessary background information on cellular communication and the commonly found smartphone architecture to understand the details of the work. Specific details and related work is covered later in the individual chapters.

First, we briefly introduce the cellular network components and some parts of cellular signaling. This is to get a rough overview of how cellular networks operate and the kind of infrastructure components mobile network operators (MNOs) have deployed. We focus on the *global system for mobile communication* (GSM) throughout this work. Second, we give a short introduction to the Short Message Service. The Short Message Service is an essential part of cellular communication and is covered in depth in this work. Third, we provide some details on the common hardware design that is found in almost every modern smartphone today. The common design is leveraged throughout this work and thus introduced here.

2.1 Cellular Network Architecture

The basic architecture of cellular networks are described in [73]. Figure 2.1 shows a simplified version of a cellular network based on [73]. We only included the network components that are important in the scope of this work. The figure leans towards a GSM network but is very similar to a 3G network at least for the scope of this work.

The architecture of a cellular network is separated into three logical subsystems. The *radio subsystem* (RSS), the *networking and switching subsystem* (NSS), and the *operation subsystem* (OSS). For packet-data service a fourth subsystem the *general packet radio service* (GPRS) is added. Below we will briefly introduce each subsystem. We vary the details depending on the need for understanding this work.

2 Background

The *radio subsystem* (RSS) comprises all radio specific entities of a cellular network. First, the base transceiver station (BTS) that basically is the radio equipment such as the actual transceiver and the antennas. Second, the base station controller (BSC) that manages one to multiple BTSs. The BSC handles tasks such as channel allocation, paging of mobile devices, and handover between BTSs. In figure 2.1 we show a compressed version of the BTS and BSC as the base station (BS). The third part of the radio subsystem is the mobile station (MS) the actual mobile phone. The mobile station comprises all the necessary hardware and software to communicate with the cellular network. A little more detail of the inner workings of a modern smartphone is described in Section 2.4.

The *networking and switching subsystem* (NSS) is the core of the cellular network. This subsystem connects the radio subsystem with the Public Switched Telephone Network (PSTN) and other necessary systems. Its duties are localization of mobile stations in the network, accounting, and charging. The two most important components of the NSS are the following. First, the mobile switching center (MSC) is the connection between the BTSs and the fixed line network. A MSC connects to multiple BSCs and interconnects to other MSCs. Second, the home location register (HLR) is the central user database of a cellular network. The HLR stores all information about a mobile subscriber from charging records and call forwarding settings to his current network location. The HLR is one of the most important components in a mobile network. There is a third component in the NSS, the visitor location register (VLR). The VLR is more or less a cache of the HLR to reduce load on the central HLR.

The *operation subsystem* (OSS) hosts some components for operating and maintaining a cellular network. It contains systems such as the authentication center (AuC) which contains the credentials of the mobile subscribers. In actual networks the AuC is part of the HLR.

The *general packet radio service* (GPRS) subsystem hosts the components necessary for packet-data access. The GPRS subsystem consist of two major entities. First, the Gateway GPRS Support Node (GGSN) which is the component that connects the GPRS network to the packet-data/IP network. Its duties are routing, address conversions, and providing tunnels for the actual data transport. The GGSN is the gateway to the public network such as the Internet. Second, the Serving GPRS Support Node (SGSN) which is responsible for authentication, billing, and mobility management (keeping track of the mobile stations that use GPRS). The GPRS subsystem interfaces with the MSC and HLR.

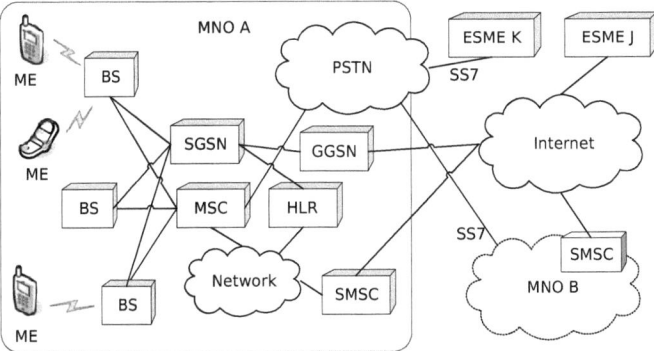

Figure 2.1: The basic setup of a cellular network.

2.2 Cellular Signaling

Signaling traffic generated by the Mobile Equipment (ME) is sent to the MSC and HLR in case of voice calls, SMS, and updating account settings (such as call-forwarding). Packet-data related signaling is mainly directed towards the SGSN, the GGSN, and of course the HLR. More details on cellular signaling can be found in [73].

Packet Data Protocol (PDP) setup in order to establish IP connectivity is a complex process. When a ME wishes to establish a PDP context it sends a GPRS-attach message to the SGSN. The SGSN authenticates the ME using the HLR. Next, the PDP context is established and stored at the SGSN and GGSN. This includes records and parameters for billing, quality of service information, and the IP address assigned to the specific PDP context. Maintenance and distribution of the PDP context information across the different network components is a costly process as it involves many components across the cellular network.

2 Background

2.3 The Short Message Service

The Short Message Service (SMS) [6] is one of the basic services of the mobile phone network. SMS is used for many different purposes besides text messaging and thus is one of the key components of cellular networks. As SMS plays such an important role it is studied heavily throughout this work.

The Short Message Service is a store and forward system. Messages sent to and from a mobile phone are first sent to an intermediate component in the mobile phone operators network. This component is called the Short Message Service Center (SMSC). After receiving a message, the SMSC forwards the message to another SMSC (in case of inter-operator messages or an operator with multiple SMSCs) or if the receiving phone is handled by the same SMSC, it delivers the message to the recipient without invoking another party. SMS messages can also be sent from entities other then mobile phones. These entities are called External Short Message Entity (EMSE). Internet-based SMS services use such EMSEs to send messages to mobile subscribers.

The format of an SMS differs depending on whether the message is `Mobile Originated (MO)` or `Mobile Terminated (MT)`. This is mapped to the two formats `SMS_SUBMIT (MO)` and `SMS_DELIVER (MT)`. In a typical GSM network, as shown in Figure 2.1, an SMS message that is sent from a mobile device is transferred Over-the-Air to the Base Station (BS) of an operator in `SMS_SUBMIT` format. To reduce the complexity in our example cellular network we define that a BS always consists of a Base Transceiver Station (BTS) and a Base Station Controller (BSC). Every BS interacts with a Mobile Switching Center (MSC), which acts as the central entity handling traffic within the network. The MSC relays the SMS message to the responsible SMSC.

The details of the SMS message formats and SMS delivery are covered in Chapters 3 and 4 where we analyze client side implementations of the Short Message Service.

2.4 Smartphone Architecture

Modern smartphones consist of two individual subsystems, the application processor and the baseband processor. Together with the peripheral hardware such as the touch

2.4 Smartphone Architecture

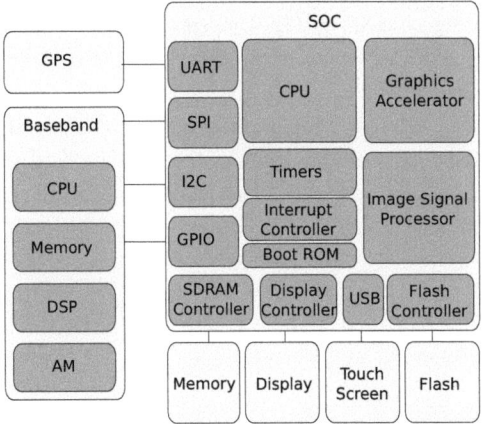

Figure 2.2: The basic design of a modern smartphone.

screen, audio input and output, and the GPS receiver these two systems form the actual smartphone. Figure 2.2 depicts the conceptual system design of a smartphone. This common design can be found across almost all chipset manufacturers, i.e. [72, 79]. This is how an iPhone, Android, and Windows Mobile device look on the inside.

The application processor usually comes in form of a System on a Chip (SoC) design. The CPU and many of the controllers for connected peripherals shown in Figure 2.2 are included on the same chip. The application processor runs the smartphone OS such as Android or iOS and all the applications (e.g., email client or telephone).

The baseband (cellular modem) processor is the communication interface to the cellular network. It consists of a general purpose CPU, a Digital Signal Processor (DSP), and the necessary radio components such as a signal amplifier. The baseband processor runs a specialized real-time operating system. Baseband chipsets are a highly specialized field since they have to be certified by multiple institutions before they are allowed to operate on public cellular networks. Because the process of development and certification is very costly, there are only a few baseband manufacturers [77].

The application processor and the baseband processor are connected at few points. This allows for better flexibility for the various phone manufacturers. The connections are for digital audio input and output (voice calls) and for controlling the baseband's functionality. The control channel is conceptually a serial connection that can be imple-

17

2 Background

mented using buses such as SPI or USB. Through this serial connection, the application processor uses an extended version of the Hayes[1] *Attention* (AT) command set [5] to interact with the baseband.

In Section 3.4 we will further explain the communication that takes place between the baseband and the application processor in the case when the phone receives a SMS messages from the cellular network. Chapter 6 deals with the commands that are emitted from the application processor to the baseband.

---

[1] Hayes Microcomputer Products initially developed the AT command set, the company does not exist anymore. http://en.wikipedia.org/wiki/Hayes_Microcomputer_Products

3 Security Analysis of SMS Clients on Smartphones

We start our investigation by analyzing the communication between the cellular modem and the application processor. Based on the results of this analysis we design a vulnerability analysis method for SMS clients of smartphones. Our resulting tool impersonates the cellular modem to emulate messages from the cellular network.

3.1 Introduction

The Short Message Service (SMS) is the most popular secondary service on mobile phones beloved by both the users and the service providers for its ease of use and for the generated revenue, respectively. Besides the use for simple text messaging, the Short Message Service has many applications on a mobile phone. It is used as a control channel for services like voice mail where it is used to notify the user about new messages. Another SMS-based service is remote over-the-air (OTA) phone configuration. SMS further is used as a transport for the Wireless Application Protocol (WAP).

The Short Message Service in all its functionality is complex and therefore security issues based on implementation faults are common. In the past years SMS-based security issues for almost every mobile phone platform were known. Furthermore, no possibility exists to firewall or filter SMS messages, therefore, SMS-based attacks are hard to prevent especially since user interaction is not required. Therefore, it is necessary to develop techniques and tools to analyze and improve the security of SMS-implementations and SMS-based applications.

In this chapter we present a novel approach to the vulnerability analysis of SMS-implementations on smartphones. To the best of our knowledge, no attempt has been made before to analyze and test Short Message Service implementations and SMS-based applications in a methodical way. We believe that the main reason for this situation is that SMS testing would be very cost intensive since SMS messages would have to be sent to the tested phones in large quantities.

The analysis of SMS-implementations on smartphones is difficult for several reasons besides the cost factor. The reasons all tail from the fact that SMS messages are delivered through infrastructure controlled by an operator and thus is outside the control of the researchers who are conducting the vulnerability analysis. One problem is the uncertainty of whether a message is delivered to the target in its original form. This is because mobile phone operators have the ability to filter and modify short messages during delivery. Also, it is possible that the operator might not filter messages on purpose but might use equipment that can not handle certain messages. Second, SMS is an unreliable service, meaning messages can be delayed or discarded for no deterministic reason. This makes the testing very time-consuming and hard to reproduce.

We addressed these problems by removing the need for a mobile phone network all together through injecting short messages locally into the smartphone. Injection is done in software only and requires only application level access to the smartphone. The injection is taking place below the mobile telephony software stack and therefore we are able to analyze and test all SMS-based services that are implemented in the mobile telephony software stack.

The vulnerability analysis itself was conducted using fuzzing. In this work, we present the possibilities for fuzzing-based testing of SMS-implementations. Further, we present our testing methodology and the tools we have developed in the process. To show that our approach is generic we implemented and tested our analysis framework for three different smartphone platforms.

So far we have found several flaws in the tested SMS-implementations, some of which can be exploited for Denial-of-Service attacks. One particular vulnerability allows us to disconnect a device from the mobile phone network through crashing the telephony application, leaving the phone in state where it cannot receive calls.

Contributions of this chapter are the following:

- We introduce a novel method to test SMS-implementations that circumvents filters and any other restrictions that might be put in place by a mobile phone service operator. It further prevents the provider from easily detecting that testing is taking place. Furthermore, it allows analysis of services and applications built on top of SMS such as WAP.

- We developed a testing framework that allows one to perform SMS vulnerability analysis at high speeds and without costs.

- We developed a tool that performs security testing of SMS-implementations through fuzzing. The tool found a number of previously unknown vulnerabilities.

Chapter Organization

The rest of this chapter is structured as follows: Section 3.2 presents related work. Section 3.3 describes the Short Message Service message format. Section 3.4 describes in great detail how SMS messages are received and handled on a smartphone. Section 3.5 describes our SMS injection framework. In Section 3.6 we present our fuzzing tools, the methodology, the results of our fuzzing approach, and the possible attacks based on the results. In Section 3.7 we briefly conclude.

3.2 Related Work

Previous research in the area of SMS security can be divided into two areas. The first area consists of research that investigated protocols that facilitate SMS as a transport such as WAP and MMS. Here the creators of the PROTOS [66] testing suite implemented test cases for WAP implementations. In [63] the authors build a framework for analyzing the security of MMS client implementations.

The second area of research conducted in [26] focused around the possibility of mobile phone service disruption based on the ability to send excessive amounts of short messages from the Internet to individuals and groups of people in a certain area.

In the past, SMS bugs [22, 48, 93] were found by accident rather than through thorough testing. One notable example of this kind of bug discovered is the Curse of Silence [27] bug which existed in most of Nokia's Symbian S60-based smartphones. The bug consisted of a single malformed SMS message that, upon reception, prevented further SMS messages from being displayed to the user. The work presented in this chapter provides a method for conducting thorough security analysis of SMS-implementations without the burden of services fees.

Name	Bytes	Purpose
SMSC	variable	SMSC address
DELIVER	1	Message flags
Sender	variable	Sender address
PID	1	Protocol ID
DCS	1	Data Coding Scheme
SCTS	7	Time Stamp
UDL	1	User Data Length
UD	variable	User Data

Figure 3.1: SMS_DELIVER Message Format

3.3 The SMS_DELIVER Format

The SMS_DELIVER format is used for messages sent from the SMSC to the mobile phone. Since our testing method is based on local message injection that replicates an incoming message, we are interested in the SMS_DELIVER format.

An SMS_DELIVER message consists of the fields shown in Figure 3.1. The format is simplified since our main fuzzing targets are the Protocol ID, the Data Coding Scheme, and the User Data fields. Other fields such as the User Data Length and the DELIVER flags will be set to corresponding values in order to create valid SMS_DELIVER messages.

The User Data Header

The User Data Header (UDH) provides the means to add control information to an SMS message in addition to the actual message payload or text. The existence of a User Data Header is indicated through the User Data Header Indication (UDHI) flag in the DELIVER field of an SMS_DELIVER message. If the flag is set, the header is present in the User Data of the message. The User Data Header consists of the User Data Header Length (UDHL), followed by one or multiple headers. The format for a single User Data Header is shown in Figure 3.2.

Field	Bytes
Information Element (IEI)	1
Information Element Data Length (IEDL)	1
Information Element Data (IED)	variable

Figure 3.2: The User Data Header (UDH)

3.4 Mobile Phone Side SMS Delivery

Most current smartphones are composed of two processors. The main CPU, called the application processor, is the processor that executes the smartphone operating system and the user applications such as the mobile telephony and the PIM applications. The second CPU runs a specialized real time operating system that controls the mobile phone interface and is called the modem. The modem handles all communication with the mobile phone network and provides a control interface to the application processor.

Logically the application processor and the modem communicate through one or multiple serial lines. The mobile telephony software stack running on the application processor and communicates with the modem through a text-command-based interface using a serial line interface provided by the operating system running on the application processor. The physical connection between the application processor and the modem solely depends on the busses and interfaces offered by both sides but is irrelevant for our method.

The modems of our test devices (the iPhone, the HTC G1 Android, and the HTC Touch 3G Windows Mobile) are controlled through the GSM AT command set [28]. The GSM AT commands are used to control every aspect of the mobile phone network interface, from network registration, call control and SMS delivery to packet-based data connectivity.

3.4.1 The Telephony Stack

The telephony stack is the software component that handles all aspects of the communication between the application processor and the modem. The lowest layer in a telephony stack usually is a multiplexing layer to allow multiple applications to access the modem at the same time. The multiplexing layer also is the instance that translates API-calls to AT commands and AT result codes to status messages. The applications

3 Security Analysis of SMS Clients on Smartphones

to allow the user to place and answer phone calls and to read and write short messages exist on top of the multiplexing layer.

3.4.2 SMS Delivery

Short messages are delivered through unsolicited AT command result codes issued by the modem to the application processor. The result code consists of two lines of ASCII text. The first line contains the result code and the number of bytes that follow on the second line. The number of bytes is given as the number of octets after the hexadecimal to binary conversion. The second line contains the entire SMS message in hexadecimal representation. Figure 3.3 shows an example of an incoming SMS message using the CMT result code which is used for SMS delivery on all of our test devices. Upon reception of the message the application processor usually has to acknowledge the reception by issuing a specific AT command to the modem. All interaction to the point of acknowledging the reception of the CMT result is handled by the multiplexing layer of the telephony stack.

```
+CMT: ,22
07916163838450F84404D0110020009030329
02181000704010200088000
```

Figure 3.3: Unsolicited AT result code that indicates the reception of an SMS message

3.4.3 The Stacks of our Test Devices

We will shortly describe the parts of the telephony stack that are relevant for SMS handling on each of our test platforms.

iOS/iPhone

On the iPhone, the telephony stack mainly consists of one application binary called CommCenter. CommCenter communicates directly with the modem using a number of serial lines of which two are used for AT commands related to SMS transfers. It handles incoming SMS messages by itself without invoking any other process, besides when the device notifies the user about a newly arrived message after storing it in the SMS database. The user SMS application is only used for reading SMS messages stored in the database and for composing new messages and does not itself directly communicate with the modem.

Android

On the Android platform the telephony stack consists of the radio interface layer (RIL) that takes the role of the multiplexing layer described above. The RIL is a single daemon running on the device and communicates with the modem through a single serial line. On top of the RIL daemon, the Android phone application (com.android.phone) handles the communication with the mobile phone network. The phone application receives incoming SMS messages and forwards them to the SMS and MMS application (com.android.mms).

Windows Mobile

In Windows Mobile, the telephony stack is quite a bit larger and more distributed compared with the iPhone and the Android telephony stacks. The parts relevant to SMS are: the SmsRouter library (Sms_Providers.dll) and the tmail.exe binary. The tmail.exe binary is the SMS and MMS application that provides a user interface for reading and composing SMS messages. Other components such as the WAP PushRouter sit on top of the SmsRouter.

3.5 SMS Injection

Based on the results of our analysis on how SMS messages are delivered to the application layer, we designed our SMS injection framework.

Our method for SMS injection is based on adding a layer between the serial lines and the multiplexer (the lowest layer of the telephony stack). We call this new layer `the injector`. The purpose of the injector is to perform a man-in-the-middle attack on the communication between the modem and the telephony stack. The basic functionality of the injector is to read commands from the multiplexer and forward them to the modem and in return read back the results from the modem and forward them to the multiplexer.

To inject an SMS message into the application layer, the injector generates a new CMT result and sends it to the multiplexer just as it would forward a real SMS message from the modem. It further handles the acknowledgement commands sent by the multiplexer. Figure 3.4 shows the logical model of our injection framework.

We implemented our injection framework for our three test platforms. We believe that our approach for message injection can be easily ported to other smartphone platforms if these allow application level access to the serial lines of the modem or the ability to replace or add an additional driver that provides the serial line interface.

We noticed several positive side effects of our framework, some of which can be used to further improve the analysis process. First of all, we can monitor and log all SMS messages being sent and received. This ability can be used to analyze proprietary protocols based on SMS, such as the iPhone's visual voice mail. The ability to monitor all AT commands and responses between the telephony stack and the modem provides an additional source of feedback while conducting various tests. On the iPhone, for example, messages are not acknowledged in a proper way if these contain unsupported features.

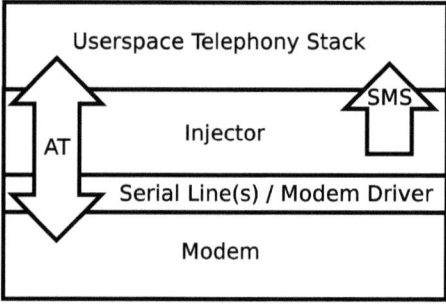

Figure 3.4: Logical model of our injector framework

3.5.1 The Injection Framework

Below we will briefly describe the implementation issues of the injection framework for each of our target platforms. Every implementation of the framework opens TCP port 4223 on all network interfaces in order to receive the SMS messages that should be injected. This network based approach gives us a high degree of flexibility for implementing our testing tools independent from the tested platform.

So far we are able to install our injection framework on all the test targets and con-

3.5 SMS Injection

tinue to use them as if the injection framework was not installed, therefore giving us high degree of confidence in our approach.

iOS/iPhone

On the iPhone, SMS messages are handled by the CommCenter process. The interface for CommCenter consists of sixteen virtual serial lines, /dev/dlci.h5-baseband. [0-15] and /dev/dlci.spi-baseband.[0-15] on the 2G and the 3G iPhone, respectively.

The implementation of our injection framework for the iPhone is separated into two parts, a library and a daemon. The library is injected into the CommCenter process through library pre-loading. The library intercepts the open(2) function from the standard C library. Our version of open checks for access to the two serial lines used for AT commands. If the respective files are opened the library replaces the file descriptor with one connected to our daemon. The corresponding device files are the serial lines 3 and 4 on the 2G and 3G iPhones. The library's only function is to redirect the serial lines to the daemon. The daemon implements the actual message injection and log functionality.

Android

The implementation for the Android platform consists of just a single daemon. The daemon talks directly to the serial line device connected to the modem and emulates a new serial device through creation of a virtual terminal.

The injection framework is installed in three steps. First, the actual serial line device is renamed from /dev/smd0 to /dev/smd0real. Second, the daemon is started, opens /dev/smd0real and creates the emulated serial device by creating a TTY named /dev/smd0. In the third step, the RIL process (/system/bin/rild), is restarted by sending it the TERM signal. Upon restart, rild opens the emulated serial line and from there on will talk to our daemon instead of the modem.

Windows Mobile

The Windows Mobile version of our injection framework is based on the simple log-driver [40] written by Willem Hengeveld. The original log-driver was designed for logging all AT communication between the user space process and the modem. We added the injection and state tracking functionality. To do this, we had to modify the driver quite a bit in order to have it listen on the TCP port to connect our test tools. The driver replaces the original serial driver and provides the same interface the original driver had and loads the original driver in order to communicate with the modem. The driver is installed through modifying several keys of the Windows Mobile registry at: /HKEY_LOCAL_MACHINE/Drivers/BuiltIn/SMD0. The most important change is the name of the Dynamic Link Library (DLL) that provides the driver for the interface, whose key is named Dll. Its original value is smd_com.dll.

3.6 Fuzzing SMS Implementations

Fuzzing is one of the easiest and most efficient ways to find implementation vulnerabilities. With this framework, we are able to quickly inject fuzzed SMS messages into the telephony stack by sending them over the listening TCP port. In general, there are three basic steps in fuzzing. The first is test generation. The second is delivering the test cases to the application, and the final step is application monitoring. All of these steps are important to find vulnerabilities with fuzzing.

3.6.1 Fuzzing Test Cases

We took a couple of approaches to generating the fuzzed SMS messages. One was to write our own Python library which generated the test cases while the other was to use the Sulley [12] fuzzing framework. In either case, the most important part was to express a large number of different types of SMS messages. Below are some examples of the types of messages that we fuzzed.

Basic SMS Messages As from Figure 3.1, we fuzzed various fields in a standard SMS message including elements such as the sender address, the user data (or message), and the various flags.

Basic UDH Messages As seen in Figure 3.5, we fuzzed various fields in the UDH header. This included the UDH information element and UDH data.

Concatenated SMS Messages Concatenation provides the means to compose SMS messages that exceed the 140 byte (160 7-bit character) limitation. Concatenation is

3.6 Fuzzing SMS Implementations

IED Byte Index	Purpose
0	ID (same for all chunks)
1	Number of Chunks
2	Chunk Index

Figure 3.5: The UDH for SMS Concatenation

IED Byte Index	Purpose
0 - 1	Destination Port (16bit)
2 - 3	Source Port (16bit)

Figure 3.6: The UDH for SMS Port Addressing

achieved through the User Data Header type 0 as specified in [6]. The concatenation header consist of five bytes, the type (IEI), the length (IEDL), and three bytes of header data (IED) as seen in Figure 3.5. By fuzzing these fields we force messages to arrive out of order or not at all, as well as sending large payloads.

UDH Port Scanning SMS applications can register to receive data on UDH ports, analogous to the way TCP and UDP applications can do so. Without reverse engineering, it is impossible to know exactly what ports a particular mobile OS will have applications listening on. We send large amounts of (unformatted) data to each port. The structure of the UDH destined for particular applications of designated ports is indicated in Figure 3.6.

Visual voice mail (iPhone only) When a visual voice mail arrives, an SMS message arrives on port 5499 that contains a URL in which the device can receive the actual voice mail audio file. This URL is only accessible on the interface that connects to the AT&T network, and will not connect to a generic URL on the Internet. The URL is clearly to a web application that has variables encoded in the URL. We fuzz the format of this URL.

3.6.2 Delivery

Once the test cases are generated, they need to be delivered to the appropriate application. In this case, due to the way we have designed the testing framework, it is possible to simply send them to a listening TCP port. All of this work is designed to make it easy to deliver the test cases.

3.6.3 Monitoring

It does no good to generate and send fuzzed test cases if you do not know when a problem occurs. Device monitoring is just as important as the other steps. Unfortunately, monitoring is device dependent. There are two important things to monitor. We need to know if a test case causes a crash. We also need to know if a test case causes a degradation of service, i.e. if the process does not crash but otherwise stops functioning properly.

On the iPhone, the crash of a process causes a crash dump file to be written to the file system compliments of Crash Reporter. This crash dump can be retrieved and analyzed to determine the kind and position of the crash. In between each fuzzed test case, a known valid test case is sent. The SMS database can be queried to ensure that this test case was received and recorded. If not, an error can be reported. In this fashion, it is possible to detect errors that do not necessarily result in a crash.

The Android development kit takes a different approach by suppling a tool called the Android Debug Bridge (ADB), this tool allows us to monitor the system log of the Android platform. If an application crashes on Android the system log will contain the required information about the crash. If a Java/Dalvik process crashes, it will contain information including the back trace of the application.

The Windows Mobile development kit on the other hand provides the tools for on-device debugging. This means Windows Mobile allows traditional fuzzing by attaching a debugger to the process being fuzzed.

3.6.4 Fuzzing Results

iOS/iPhone: Our iPhone targets were running software versions 2.2, 2.2.1, and 3.0. We discovered multiple Denial-of-Service attacks and one code execution bug. One of the flaws we discovered here is a `null pointer dereference` in the handling code of Flash-SMS messages. The flaw causes SpringBoard (the iPhone window manager)

3.7 Conclusions

to crash forcing the user to `slide to unlock` his iPhone. The crash and restart of SpringBoard blocks the phone for about 15 seconds. Second we discovered a memory corruption bug in CommCenter. When CommCenter crashes all wireless interfaces are disconnected, allowing to interrupt phone calls in process! Later we discovered that the bug allowed arbitrary code execution. To gain control over to program counter 519 SMS messages need to be sent. Although the user only sees the last message.

Android: For Android, the targets were the development firmware versions 1.0, 1.1 and 1.5. Here we found several flaws that cause an `array index out of bounds` exception. Multiple of the flaws cause `com.android.phone` to crash and thereby disconnect the phone from the mobile phone network. Additionally, if the SIM card had a PIN set the phone will be disconnected until the user enters the PIN. Thus leaving the phone disconnected from the cellular network until the user discovers the PIN entry request.

Windows Mobile: On our Windows Mobile target (a HTC Touch 3G) we discovered a bug in the HTC TouchFlo (Manila2D.exe) application. TouchFlo provides the user interface to this series of Windows Mobile phone. The bug crashes the TouchFlo application thus rendering the phone unusable. To make things worse TouchFlo stores the message in its Inbox before parsing it and thus crashes again after restarting while trying to read the Inbox. The bug was a simple format string bug that can be send from any phone. It consists of only the two characters `%n` and thus can be exploited by anyone who is able to send a simple text message.

In order to determine if a specific flaw can be exploited the particular SMS message needs to be sent over the mobile phone network. If the message is delivered to the target, and was not modified in the process, it can be utilized for an attack.

3.7 Conclusions

We presented a novel method for performing vulnerability analysis of SMS-implementation on smartphones. Our method removes the cost factor and thus enables large scale fuzz-based testing. In addition, it removes the intermediate infrastructure that otherwise would make obtaining conclusive results difficult. Removing the infrastructure further creates the possibility to discover flaws that could not haven been discovered through testing using a service providers infrastructure. Through the use of our testing tools, we identified a number of vulnerabilities that can be abused for critical Denial-of-Services attacks and in one case for arbitrary code execution. Especially the confirmation of the arbitrary code execution would not have been possible without our injection framework.

Future work will focus on porting our framework to other mobile phone platforms for testing and analyzing more SMS-implementations. We further believe that our injection framework can be used beyond the focus of fuzz-based testing, since it provides an unfiltered and cost free path for delivering SMS messages to a smartphone.

4 Security Analysis of SMS Implementations on Feature Phones

In this chapter we continue with vulnerability analysis of SMS implementations but switch the target to phones that implement the cellular modem and the applications on the same CPU. We present the resulting impact of this design on the security and reliability of the respective mobile phones.

4.1 Introduction

In recent years a lot of effort has been put into analyzing and attacking smartphones [37, 45, 57–59, 62, 63, 81], neglecting the so-called feature phones. Feature phones, mobile phones that have advanced capabilities besides voice calling and text messaging, but are not considered smartphones, make up the largest percentage of mobile devices currently deployed on mobile networks around the world. In comparison, smartphones only account for about 16% of all mobile phones [11]. The lack of security research into the far more popular feature phones is explained by the fact that smartphones share much commonality with desktop computers, and, therefore, are easier to analyze. Researchers are able to use the same or similar tools that they are already familiar with on desktop computers. Feature phones on the other hand are highly embedded systems that are closed to developers. This results in billions (there are about 4.6 billion mobile phone subscribers [9, 11]) of potentially vulnerable mobile devices out in the field, just waiting to be taken advantage of by a knowledgeable attacker.

In this chapter, we investigate the security of feature phones and the possibility for large scale attacks based on discovered vulnerabilities in these devices. We present a novel approach to the vulnerability analysis of feature phones, more specifically for their SMS client implementations. SMS is interesting because it is the feature that exists on every mobile phone. Furthermore, security issues related to SMS messaging can be exploited from almost anywhere in the world, and thus present the ideal attack vector against such devices. To the best of our knowledge, no attempt has been made before to analyze or test feature phones for security vulnerabilities.

Analyzing feature phones is difficult for several reasons. First of all, feature phones are completely closed devices that do not allow for development of native applications and do not provide debugging tools. Moreover, analyzing the part of the phone that interacts with the mobile phone network is hard since the mobile phone network between us and the target device is essentially a black box. As a consequence, analysis becomes time consuming, unreliable, and costly.

We address these problems by building our own GSM network using equipment that can be bought on the market. *We use this network not only for sending SMS messages to the phones we analyze, but also as an advanced monitoring system. The monitoring system replaces our need for debuggers and other tools that are normally required for thorough vulnerability analysis, but do not exist for feature phones.*

Vulnerability analysis was conducted using fuzzing. We chose fuzzing as the testing technique because we did not have access to source code and reverse engineering a large number of devices is not feasible. Additionally, fuzzing proved to be very efficient since this allowed us to analyze a large amount of mobile handsets with the same set of tests.

So far, we have found numerous vulnerabilities in feature phones sold by the six market leading mobile phone manufacturers. The vulnerabilities are security critical as they can remotely crash and reboot the entire target phone. In the process the mobile phone is disconnected from the mobile network, interrupting any active calls and data connections. Such bugs and attacks have existed before on the Internet, known as Ping-of-Death [49]. We believe this represents a serious threat to mobile telephony world wide.

To complete our research we further analyzed the effect of such attacks on the mobile phone core network. This resulted in two interesting findings. First, the mobile phone network can be abused to amplify our Denial-of-Service attacks. Second, by attacking mobile phones one can attack the mobile phone network itself.

The main contributions of this chapter are:

- **Vulnerability Analysis Framework for Feature Phones:** We introduce a novel method to conduct vulnerability analysis of feature phones that is based on a small GSM base transceiver station. We solve the major issue of such analysis: the monitoring for crashes and other unexpected behavior. We present multiple solutions for monitoring such devices while analyzing them. Our method furthermore shows that once a system, such as GSM, becomes partially open, the security of the entire system, including the parts that are still closed, can be analyzed and exploited.

- **Bugs Present in Most Phones:** We show that vulnerabilities exist in most mobile phones that are deployed on mobile networks around the world today. The bugs we discovered can be abused for carrying out large scale Denial-of-Service attacks.

- **Attack Impact:** We show that a small number of bugs in the most popular mobile phone brands is enough to take down a significant number of mobile phones around the world. We further show that bugs present in mobile phones can possibly be used to attack the mobile phone network infrastructure.

Chapter Organization

The rest of this chapter is structured in the following way. In Section 4.2 we discuss related work and show how our research extends previous work in this area. In Section 4.3 we explain how we selected our targets for analysis and resulting attacks. In Section 4.4 we show in great detail how to analyze feature phones for security vulnerabilities. In Section 4.5 we layout methods to use the vulnerabilities discovered for large scale attacks on mobile communication. In Section 4.6 we present methods for detecting and preventing the attacks we designed. In Section 4.7 we briefly conclude.

4.2 Related Work

Related work is separated into four parts. First, smartphone vulnerability analysis. Second, mobile and feature phone bugs, which were all found purely by accident. Third, studies on attacks against mobile phone networks. Fourth, Denial-of-Service (DoS) attacks since we are going to present a large scale mobile phone DoS attack in this chapter.

4 Security Analysis of SMS Implementations on Feature Phones

The authors of [63] built a framework for security analysis of Multimedia Messaging Service (MMS) implementations on Windows Mobile based smartphones. In Chapter 3 we conducted vulnerability analysis of Short Message Service (SMS) implementations of smartphones. Both cases used traditional techniques such as debuggers and analysis of crash dumps to catch exceptions generated during fuzzing.

Our work presented in this work is different, as we do not rely on debugging capabilities provided by the various manufacturers, which mostly do not provide such capabilities at all. Instead we use a small GSM base station to monitor and catch abnormal behavior of the phones by monitoring and analyzing radio link activity. MMS-based attacks that lead to battery exhaustion due to increasing power consumption have been studied in [70]. They utilized the fact that MMS messages use more battery resources because of GPRS and increased CPU usage. However, we did not conduct this kind of analysis since our focus was software bugs in SMS implementations.

Over the last few years a small number of bugs have been discovered by individuals. Most of them have been found by accident. To our knowledge no systematic testing has been conducted. Some examples are: the Curse-of-Silence [27] named bug for Symbian OS that prevents a phone from further receiving any SMS after receiving the *curse* SMS message. The WAP-Push vCard bug on Sony Ericsson phones [60] that caused a target phone to reboot. Some Nokia phones [93] contained a bug that could be abused to remotely crash a phone by sending it a specially crafted vCard via SMS. Some mobile phones produced by Siemens contained a bug [48] that would shutdown the phone when displaying an SMS message that contained a special character. Bugs like these fuelled our research effort since we believed that most phones contain similar bugs. A large number of similar issues in an exploit arsenal can likely be used to carry out attacks against a bigger percentage of mobile phone users around the world.

Enck et al. show in [26] that SMS messages sent over the Internet can be used to carry out a Denial-of-Service attack against mobile phone networks. The attack focused on blocking the mobile network's control channels, therefore, no more calls could be initiated. Solutions against this type of resource consumption attack are investigated in [84]. However our attacks, described in this work, are not based on attacking the radio link (the control channel) in any way. We attack the handsets directly without targeting the control channel. A study on the capabilities of mobile phone botnets [85] shows that these could be used to carry out DoS attacks against a mobile network. The attack works by overloading the Home Location Register (HLR) by triggering large amounts of state changes by zombie phones. However, in this work we show that one can achieve a similar kind of DoS attack against an operators network by disconnecting large amounts of mobile phones from the network. The difference to the botnet approach is that we do not need to have control over the zombie phones in the first place. We can remotely force them to reboot and disconnect and re-authenticate to the network and thus cause a higher load on the network core infrastructure.

Denial-of-Service attacks such as the one presented in this work have been studied in a wide area. Attacks ranging from the Web to DNS [29]. More interesting in our context are attacks that disable real-world systems and processes such as emergency services [61] (although just as a side effect) or even postal services [15].

Essentially the work presented in this work is different in many aspects. We focus on feature phones because feature phones are much more popular than smartphones. Therefore, attacks against feature phones have a larger global impact. In this work we present a security testing framework for analyzing SMS implementations of any kind of mobile phone. We used this framework to analyze feature phones of the most popular manufacturers in the world, as shown in Section 4.3. We also performed this type of analysis because it has not been done in the past, even though these devices are widely deployed.

4.3 Target Selection

To achieve maximum impact with an attack, it makes sense to target the most popular devices. We determined that feature phones are the dominant type of mobile phones. They account for 83% of the U.S. mobile market [33], smartphones in comparison just make for 16% of all mobile phones world wide [11]. We acknowledge that today smartphone sales are rising very fast, but feature phones still dominate when it comes to deployed devices in the field.

Most of the definitions of the term *feature phone* are a bit fuzzy. A loose definition of the term is: every mobile phone that is neither a dumb phone nor a smartphone is considered a feature phone. Dumb phones are phones with minimal functionality, often they only support voice calls and sending SMS messages, just basic functionality. Feature phones have less functionality than smartphones but still more than dumb phones. Feature phones have proprietary operating systems (firmware) and have additional features (thus the term feature) such as playing music, surfing the web, and running simple applications (mostly J2ME [78]). Despite this lack of functionality (compared to smartphones) they are quite popular because they are cheap and offer long battery life.

Technically interesting is the fact that feature phones are based on a single processor that implements the baseband, the applications, and user interface. Smartphones usually have a dedicated processor for the baseband. The consequence of this is that a simple bug on a feature phone may bring down the complete system.

Mobile phones are produced by many different manufacturers that all have their own OS, therefore, targeting a single one of them will not result in global effect. Since we can not simply target all mobile phone platforms we have to select the few ones that have enough market share to be of global relevance.

To determine the major mobile phone manufacturers we analyzed various market reports: World wide [10] and European [43] market share. Market shares in the United States [20] and in Germany [19]. We compiled the essence of these reports as four tables in Table 4.1.

Through this analysis we got a clear picture about the top manufacturers. These are Nokia, Samsung, LG, Sony Ericsson, and Motorola. We further chose to add Micromax [56] to the list of interesting mobile phone manufacturers because we read [82] that they are the third most popular brand of mobile phones in India.

4.3 Target Selection

(a) Germany, November 2009

Manufacturer	Market Share
Nokia	35.4%
Sony Ericsson	22.0%
Samsung	15.0%
Motorola	8.6%
Siemens	5.4%

(b) U.S.A., May 2010

Manufacturer	Market Share
Samsung	22.4%
LG	21.5%
Motorola	21.2%
RIM	8.7%
Nokia	8.1%

(c) Europe, June 2010

Manufacturer	Market Share
Nokia	32.8%
Samsung	12.5%
LG	4.1%
Sony Ericsson	3.7%
Apple	3.0%
RIM	2.4%
Others	3.0%

(d) World, for the year 2009

Manufacturer	Market Share
Nokia	38.0%
Samsung	20.0%
LG	10.0%
Sony Ericsson	5.0%
Motorola	5.0%
ZTE	4.5%
Kyocera	4.0%
RIM	3.5%
Sharp	2.6%
Apple	2.2%
Others	5.0%

Table 4.1: Mobile phone manufacturer market shares

4.4 Security Analysis of Feature Phones

Analyzing feature phones for security vulnerabilities is hard for several reasons. There is no access to source code of the OS and applications. There are no existing native-SDKs, therefore, there is no way to run native code on the device and further no access to a debugger. JTAG-based debugging is also no option since not all devices have JTAG enabled. Furthermore, deeper knowledge of the hardware and software is required in order to use JTAG debugging in a meaningful way.

Because of these reasons we choose to conduct fuzz-based testing. The testing is carried out on our own GSM network. In order to monitor for misbehavior, crashes, and to find the related bugs, we designed our own monitoring system. Throughout this section we will first describe the setup of our GSM network. Followed by the way we send SMS messages in this setup. Then we will describe our novel monitoring setup. The final part of the section will discuss test cases and the resulting bugs that were discovered throughout this work.

4.4.1 Network Setup

Since we want to send large amounts of SMS messages we decided to build our own GSM network rather than sending SMS messages over a real network. On the one hand this has the advantage of not costing any money and on the other hand we do not risk to interfere with the telecommunication networks. We want to avoid crashing the operator's network equipment by either content or quantity of SMS messages. Having our own network assures reproducible results because we have control of the entire system and are able to quickly find parameters that cause unexpected results. Analysis over a real operator network would only leave us with the possibility of guessing in many cases. In addition, the delivery of SMS messages is much faster on our small network compared to a production setup of a mobile operator.

On the hardware side we decided to use an ip.access nanoBTS [46], which is a small, fairly cheap (about 3500 Euro) GSM Base Transceiver Station (BTS) that provides an A-bis over IP interface. The A-bis interface is used to communicate between the BTS and the Base Station Controller (BSC). The BSC part of our setup is driven by OpenBSC [92]. OpenBSC is a Free Software implementation of the A-bis protocol that implements a minimal version of the BSC, Mobile Switching Center (MSC), Home Location Register (HLR), Authentication Center (AuC) and Short Message Service Center

4.4 Security Analysis of Feature Phones

(SMSC) components of a GSM network. Figure 4.1 shows a picture of our setup.

Figure 4.1: Our setup: A laptop that runs OpenBSC and the fuzzing tools, the nanoBTS, and some of the phones we analyzed.

As GSM operates on a licensed frequency spectrum we had to carry out our experiments in an Faraday cage.

Utilizing this setup we are able to send SMS messages to a mobile phone. OpenBSC allows us to either send a text message from its telnet interface to a subscriber of our choice or it processes an SMS message that it received Over-the-Air in a store and forward fashion. As we later see the existing interface is not feasible for fuzzing since we need the ability to closely control all parameters in the encoded SMS format as well as a way to inject binary payloads.

Using a mobile phone to inject SMS messages into the network is not an option as this would be very slow as we show later. Instead we built a software framework based on a modified version of OpenBSC that allows us to:

- Inject pre-encoded SMS into the phone network

- Extensive logging of fuzzing related feedback from the phone

- Logging of non-feedback events, i.e. a crash resulting in losing connection to the network

- Automatic detection of SMS that caused a certain event

4 Security Analysis of SMS Implementations on Feature Phones

- Process malformed SMS with OpenBSC
- Smart fuzzing of various SMS features
- Ability to fuzz multiple phones at once
- Sending SMS at higher rate than on a real network

Compared to a real cellular network, like the one shown in Figure 2.1, in our setup OpenBSC acts as BSC, MSC, and SMSC. During the final transmission to the destination the SMS will get converted to SMS_DELIVER, this is taken care of by OpenBSC. Our fuzzer we presented in Chapter 3 generated SMS messages in the SMS_DELIVER format since we directly fuzzed on the phone without using any kind of cellular network. Both formats are similar and no field that is subject to our fuzzing is lost. SMS_SUBMIT only contains the destination number and since SMS works in a store-and-forward fashion, the destination address is replaced with the sender number on the final transmission to the destination. SMS_DELIVER does not include the destination number but instead relies on an existing channel to the phone (after the phone has been paged). For this reason we utilize the SMS_SUBMIT format when injecting messages.

4.4.2 Sending SMS Messages

OpenBSC itself does not provide an interface to submit pre-encoded SMS messages to the network, but only an interface to submit text SMS messages that are then converted into the corresponding encoding. We added a new interface to OpenBSC that allows us to submit SMS messages directly in SMS_SUBMIT format. These messages are inserted into a database that is used by OpenBSC as part of the SMSC functionality. In our version not only the parsed SMS values are stored, but also the complete encoded message for easy reproducibility. Modifying the existing text message interface to be capable of handling binary encoded SMS messages proved to be infeasible. Messages submitted over this interface are instantly transmitted to the subscriber if he is attached to the network. This means opening a channel, initiating a data connection, sending the message and tearing down the connection. This works, but is very slow and takes about seven seconds per message. This is also the reason why we did not want to use a mobile phone to send our fuzz-messages in the first place. Our method of injecting messages is much faster. Prior to testing we use our new interface to inject thousands of messages into the SMSC database. Next, we send these messages. Ideally, this only

4.4 Security Analysis of Feature Phones

opens a channel once and sends all SMS messages (pending delivery) to the recipient and then closes the connection. This greatly improves the speed at which we can fuzz since the actual message transfer only takes about one second.

In essence we removed the sending mobile phone and replace it with a direct interface to the network. This way it was not necessary to modify the target mobile phone in any way.

4.4.3 Monitoring for Crashes

In fuzz-based testing, monitoring is one of the essential parts. Without good monitoring one will not catch any bugs.

OpenBSC itself already has an error handler that takes care of errors reported from the phone, which we modified to fit our fuzzing case. The default error handler does not differentiate between errors and is not taking the cause of an error into account. It simply stops the SMS sending process in case of an error. The only exception is a `Memory Exceeded` error, which causes OpenBSC to dispatch a signal handler to wait for an SMMA signal (released short message memory) indicating that there is enough space again.

The mobile phone as well as the MSC are usually divided into separated layers for transferring and processing a message. As shown in Figure 4.2 they consist of a Short Message Transport Layer (SM-TL), Short Message Relay Layer (SM-RL) and the Connection Sublayer (CM-Sub). The SM-TL [2] receives and relays messages that it receives from the application layer in TPDU form (Transport Protocol Data Unit). This is the original encoding form that we describe later in this chapter. The message is passed to the SM-RL to transport the TPDU to the mobile station. At this point the TPDU is encapsulated as an RPDU. As soon as a connection is established between the mobile station and the network the RPDU is transferred Over-the-Air encapsulated in a CP-DATA unit that is part of Short Message Control Protocol (SM-CP). Both sides communicate via their CM-Subs with each other. The CM-Sub on the phone side will unpack the CPDU and forward the encapsulated TPDU to the Transport Layer using an RP-DATA unit. At this point the mobile phone stack has already performed sanity checks on the content of the SMS and parsed it. The resulting reply, passed to CM-Sub, will include an acknowledgement of the SMS message and it will then be passed to the higher layers. From there it will end up in the user interface or an error message is encapsulated and sent back to the network. For our monitoring we need to log these replies carefully to observe the status of the phone.

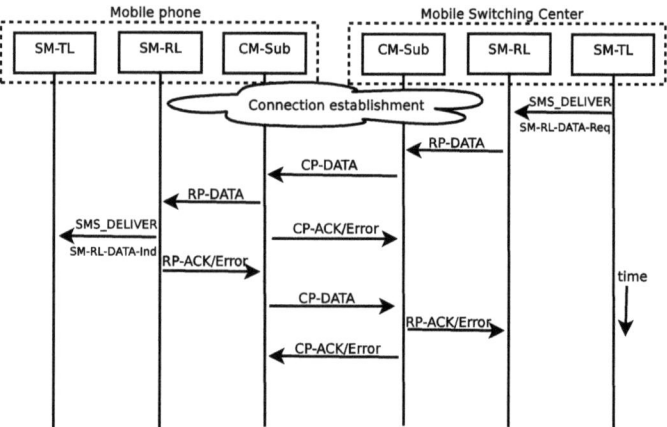

Figure 4.2: Mobile terminated SMS

From the wide variety of error messages a phone can reply to a received SMS message (defined in [3]), we observed during our fuzzing experiments that all of the tested phones either reply with a Protocol Error or Invalid Mandatory Information message in the case of a malformed message. These two responses besides the memory error have been the only errors that we observed in practice. We added code to flag such an SMS message as invalid in the database and continue delivering the next SMS that has not been flagged as invalid. OpenBSC would otherwise continue trying to retransmit the malformed SMS message and thus block further delivery for the specific recipient.

SMS messages are usually sent over a SDCCH (Stand-alone Dedicated Control Channel) or a SACCH (Slow Associated Control Channel). The details of such a channel are not important for the scope of this work. However the use of such a logical channel is an important measurement to detect mobile phone crashes. Such a channel will be established between the BTS and the phone on the start of an SMS delivery by paging the phone on a broadcast channel. As we explained earlier, we only open the channel once and send a batch of messages using this one channel. The channel related signaling between the BSC and the BTS happens over the A-bis interface over highly standardized

4.4 Security Analysis of Feature Phones

protocols. We added modifications to the A-bis *Radio Signaling Link* code of OpenBSC that allows us to check if a channel tear down happens in a usual error condition, log when this happens and which phone was previously assigned to this channel.

So while we lack possibilities to conduct traditional debugging methods on the device itself we can use the open part - OpenBSC - to do some debugging on the other end of the point-to-point connection.

The difference to traditional debugging techniques is that we are mostly limited towards noticing an error condition and monitoring the impact of such an error. We are not able to peek at register values and other software related details of the phone firmware. However, it is enough to be able to reliably detect and reproduce the error. Using this method it also possible to find code execution flaws. However exploiting them and getting to know the details about the specific behavior requires the effort of reverse engineering the firmware for a specific model. We try to avoid such a large scale test of phones but these bugs are a good base for further investigations such as reverse engineering of firmware.

In the next step we have written a script that parses the log file, evaluates it and takes actions in order to determine which SMS message caused a problem.

When delivering an SMS message to a recipient phone under the assumption that it is associated with the cell in practice three things can happen. Either the message is accepted and acknowledged, it is rejected with a reason indicating the error, or an unexpected error occurs. Such an unexpected error can be that the phone just disconnected because it crashed or due to other reasons the received message is never acknowledged. In the latter case, OpenBSC stores the SMS message in the database, increases a delivery attempt counter and tries to retransmit the SMS message when the phone associates with the cell again. For our fuzzing results this means that this method detects bugs in which the SMS message either results in a phone crash after it accepted the message or already during receiving it in which it will never be acknowledged and OpenBSC continuously tries to deliver the SMS message.

Detecting the SMS message that caused such an error condition then is fairly simple. Our script checks the error condition and if it occurred because of the loss of a channel it first looks up the database to find SMS messages that have a delivery count that is bigger or equal to one and the message is not marked as sent (meaning it was not acknowledged). In this case we can with a high probability say that the found SMS message caused the problem. If there is no message the script checks which messages have been sent in a certain time interval around the time of the log event. During our testing we decided that a one minute time interval works well enough to have a fairly small subset of candidate SMS messages that could have caused a problem. Figure 4.3 shows the logical view of our monitoring setup.

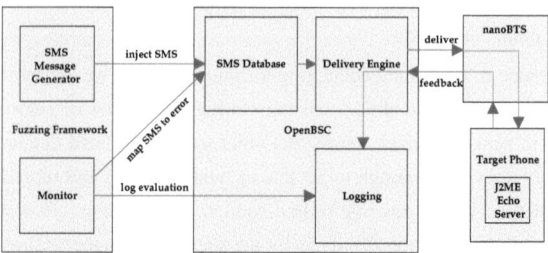

Figure 4.3: Logical view of our setup

4.4.4 Additional Monitoring Techniques

In addition to the aforementioned OpenBSC setup we have developed more methods for monitoring for abnormal behavior.

Bluetooth: Bluetooth can be used to check if a device crashes or hangs. Our monitor script connects to the device using a Bluetooth virtual serial connection (RFCOMM) by connecting to the RFCOMM channel for the phone's dial-up service. The script calls recv(2) and blocks since the client normally is supposed to send data to the phone. When the phone crashes or hangs, the physical Bluetooth connection is interrupted and recv(2) returns, thus signaling us that something went wrong.

J2ME: Almost every modern feature phone supports J2ME [78] and this is providing us with the only way to do measurements on the phone since they do not run native applications. Applications running on the mobile phone can register a handler in an SMS registry similar to binding an application to a TCP/UDP port. SMS can make use of a User Data Header [2] (UDH) that indicates that a certain SMS message is addressed to a specific SMS-port. When the phone receives a message this header field will be parsed and the message is forwarded to the application registered for this port.

4.4 Security Analysis of Feature Phones

Our J2ME application that is installed to the fuzzed phone registers to a specific port and receives SMS messages on it. For each chunk of fuzzed SMS messages we inject a valid message that is addressed to this port. The application then replies with an SMS message back to a special number that is not assigned to a phone. Figure 4.3 shows this as the J2ME echo server. The message is just saved to the SMS database. This allows us to easily lookup the count of SMS messages for this special number in the database and check if it increased or not. If not, it is very likely that some odd behavior was triggered. This kind of monitoring is useful to identify bugs that block the phone from processing received messages such as those described in [27].

4.4.5 SMS_SUBMIT Encoding

The SMS_SUBMIT format as defined in [2] consists of a number of bit and byte fields, the destination address, and the message payload. Below we briefly describe the parts the are important for our analysis. Figure 4.4 shows the structure of an SMS_SUBMIT message.

Field	Size
TP-Message-Type-Indicator	2 bit
TP-Reject-Duplicates	1 bit
TP-Validity-Period-Format	2 bit
TP-Status-Report-Request	1 bit
TP-User-Data-Header-Indicator	1 bit
TP-Reply-Path	1 bit
TP-Message-Reference	integer
TP-Destination-Address	2-12 byte
TP-Protocol-Identifier	1 byte
TP-Data-Coding-Scheme	1 byte
TP-Validity-Period	1 byte/7 byte
TP-User-Data-Length	integer
TP-User-Data	depends on DCS/UDL

Figure 4.4: Format of the SMS_SUBMIT PDU

TP-Protocol-Identifier (1 octet) describes the type of messaging service being used. This references to a higher layer protocol or *telematic interworking* being used. While

4 Security Analysis of SMS Implementations on Feature Phones

this is included in the specifications, we believe that these interworkings are mostly legacy support and not in use these days. This makes it an interesting target to study unusual behavior.

TP-Data-Coding-Scheme (1 octet) as described in [1] indicates the message class and the alphabet that is used to encode the *TP-User-Data* (the message payload). This can be either the default 7 bit, 8 bit or 16 bit alphabet and a reserved value.

The *TP-User-Data* field together with the *TP-Protocol-Identifier* and the *TP-Data-Coding-Scheme* are the main targets for fuzzing. The receiving phone parses and displays the message based on this information.

However these fields are not enough to cover the complete range of possible SMS features. If the *TP-User-Data-Header-Indicator* bit (one of the earlier mentioned bit fields) is set this indicates that *TP-User-Data* includes a UDH.

The UDH is used to provide additional control information like headers in IP packets. It can hold multiple so called Information Elements [6] (IEI), for example elements for port addressing, message concatenation, text formatting and many more. IEIs are represented in a simple type-length-value format. Figure 4.5 shows an example UDH with multiple IEIs.

Field	Size
UDHL	1 byte
IEI_1	1 byte
$IEIDL_1$	1 byte
$IEID_1$	n bytes
...	
IEI_n	1 byte
$IEIDL_n$	1 byte
$IEID_n$	n bytes

Figure 4.5: The User Data Header

4.4.6 Fuzzing Test-cases

We have implemented a subset of the SMS specification as a Python library to create SMS PDUs (Protocol Data Unit) and used this to develop a variety of fuzzers. This includes fuzzers for vCard, vCalendar, Extended Messaging Service, multipart, SIM-Data-Download, WAP push service indication, flash SMS, MMS indication, UDH, simple text messages and various others fuzzing only single fields that are part of a specific SMS feature. Some of these features can also be combined. For example most of the features can either consist of single SMS message or be part of a multipart sequence by adding the corresponding multipart UDH.

For the scope of this work we focused on fuzzing multipart, MMS indication (WAP push), simple text, flash SMS, and simple text messages with protocol ID/data coding scheme combinations. These test cases cover a wide variety of different SMS features.

Multipart: SMS originally was designed to send up to 140 bytes of user data. Due to 7-bit encoding it is possible to send up to 160 bytes. However various SMS features rely on the possibility to send more data, e.g. binary encoded data. Multipart SMS allow this by splitting payload across a number of SMS messages. This is achieved by using a multipart UDH chunk (IEI: 0, length: 3). This UDH chunk comprises three one byte values. The first byte encodes a reference number that should be random and the same in all message parts that belong to the same multipart sequence. Based on this value the phone is later able to reassemble the message. The second byte indicates the number of parts in the sequence and the last byte specifies the current chunk ID. By fuzzing these three values we were mainly looking for abnormal behavior related to combinations of the current chunk ID and the number of chunks in a sequence. For example missing chunk and chunk IDs higher than the number of total chunks.

MMS indication: When a subscriber receives an MMS (Multimedia Messaging Service) message an MMS notification indication message [89] is sent to him. This MMS indication is in fact a binary encoded WAP-push message sent via SMS. The notification contains multiple variable length fields for subject, transaction ID and sender name. There are no length fields for these values. They are simple zero terminated hex strings. An MMS indication message can also consist of multipart sequences. Therefore, our fuzzing target were the variable length field values included in the message seeking for classic issues like buffer overflow vulnerabilities.

Simple text: Implementations of decoders for simple 7 bit encoded SMS often work with a GSM alphabet represented for example with an array. The decoder first needs to unpack the 7 bit encoded values and convert them to bytes. After this step it can lookup the character values in the GSM alphabet table. Our fuzzers mixed valid 7 bit sequences with invalid encodings that would result in no corresponding array index. This could trigger all kinds of implementation bugs but most noteworthy out of bounds access resulting in null pointer exceptions and the like.

TP-Protocol-Identifier/TP-Data-Coding-Scheme: The combination of both of these fields defines how the message is displayed and treated on the phone. Both of these fields are one byte values and also cover several rather unpopular features and reserved values. With fuzzing combinations of these values with random lengths of user data payload we were aiming for odd behavior and bugs in code paths that are seldom used by normal SMS traffic.

Flash SMS: Flash messages are directly displayed on the phone without any user interaction and the user can optionally save the message to the phone memory. Our observations made it clear that often the code that renders the flash SMS message on the display is not the same as the one that displays a normal message from the menu. Therefore, it can be prone to the same implementation flaws as simple text messages. Additionally, flash SMS can consist of multipart chunks and there are several combinations of TP-Protocol-Identifier and TP-Data-Coding-Scheme that cause the phone to display the SMS as flash message. Our flash SMS fuzzers aim to cover a combination of all of the above possible implementation weaknesses.

4.4.7 Fuzzing Trial

After each fuzzing-test-run we evaluate the log generated by our monitoring script. All of the bugs described later in this chapter were triggered by one or very few SMS messages and reproducing problems from log entries was rarely problematic. However, during our fuzzing studies we stumbled across various forms of strange behavior. Problems we faced included non-standard conforming message replies and various kinds of weird behavior. Some phones were not properly reporting memory exhaustion. Others did not notice free memory until a reboot. Some did not display a received SMS message on the user interface which made it hard to tell if the phone accepted a message or silently discarded it on the phone. Almost every phone we fuzzed needed a hard reset at some point because it became simply unusable for unknown reason, the mass of messages or a specific SMS needed to be deleted from the SIM card using another phone. One of the biggest issues we came across was that very few manufacturers' hard reset actually restored the phone to an initial factory state. From what we know this is done as a feature for customers in order to ensure no personal data is lost. The behavior also differed between phones of the same manufacturer. When testing a bug on the Samsung B5310 it was always sufficient to remove the offending SMS message from the phone's SIM card while the Samsung S5230 needed an additional hard reset. Understanding such issues proved to be extremely time-consuming. However, it is worth noting that purging a phone of all personal information can prove to be nearly impossible for a user. This can become an issue whenever a user plans to sell a used handset to a third party.

4.4.8 Results

During our fuzz-testing we discovered quite a few bugs that lead to security vulnerabilities. The bugs mostly lead to phones crashing and rebooting, which disconnected the phones from the mobile network and interrupted ongoing voice calls and data connections. Our testing even resulted in two *bricked phones* that could no longer be reset and brought back into working order. We did not investigate the bricking in-depth because this would have gotten quite costly. Furthermore, some of the phones crash during the process of receiving the SMS message, and, therefore, fail to acknowledge the message thus causing re-transmission of the SMS message by the network.

Below we present some of the bugs we discovered on each platform. In most cases we fuzzed only one phone from each platform and later only verified the bugs on other phones we had access to. This is expected because most manufacturers base their entire product line on a single software platform. Only customizing options such as the user interface depending on the hardware of a specific device.

We reported all bugs to the manufacturers including full PDUs in order to verify and reproduce them. The feedback we received indicates that the bugs are present in most of their products based on their feature phone platforms. So far we have not received any information about fixes or updates.

4 Security Analysis of SMS Implementations on Feature Phones

Nokia S40: On our test devices `6300, 6233, 6131 NFC, 3110c` we found a bug in the flash SMS implementation. The phones run different versions of the S40 operating system, the oldest of which was over 3 years older than the newest. The manufacturer confirmed that this bug is present in almost all of their S40 phones. By sending a certain flash SMS the phone crashes and triggers the "Nokia white-screen-of-death". This also results in the phone disconnecting and re-connecting to the mobile phone network. Most notably, the SMS actually never reaches the mobile phone. The phone will crash before it can fully process and acknowledge the message. On the one hand this has the side effect that the GSM network performs a Denial-of-Service attack for free as it continuously tries to transmit the message to the phone. On the other hand this has a side effect on the phone since there seems to be a watchdog in place that is monitoring such crashes. This watchdog shuts down the phone after 3 to 5 crashes depending on the delay between the crashes.

Sony Ericsson: Our test devices `W800i, W810i, W890i, Aino` running OSE have a problem similar to the Nokia phones. When combining certain payload lengths together with a specific protocol identifier value it is possible to knock the phone off the network. In this case there is no watchdog, but one SMS message is enough to force a reboot of the phone. As in the case of the Nokia bug, this SMS message will never be acknowledged by the phone. To get an idea on how wide spread the problem is, we investigated the age of the devices and found that the oldest phone (W800i) is from 2005 while the newest phone (Aino) is from late 2009.

LG: Our `LG GM360` seems to do insufficient bounds checking when parsing an MMS indication message. This allows us to construct an MMS indication SMS message containing long strings that span over three or more sms. This crashes the phone and thus forces an unexpected reboot when receiving the message or as well when trying to open the SMS message on the phone.

Motorola: As aforementioned, SMS supports *telematic interworking* with other network types. By sending one SMS message that specifies an Internet electronic mail interworking combined with certain characters in the payload it is possible to knock the phone off the mobile network. Upon receiving the message the phone shows a flashing white screen similar to the one shown by the Nokia phones. The phone does not completely reboot; instead it simply restarts the user interface and reconnects to the network. This process takes a few seconds and depending on the payload it is possible to achieve this twice in a row with one message. We verified this on the `Razr, Rokr`, and the `SVLR L7` – older, but extremely popular devices. The devices span 3+ years, providing us with confidence that the bug is present in their entire platform.

Samsung: Multipart UDH chunks are commonly used for payloads that span over multiple SMS messages. The header chunk for multipart messages is simple.

Our Samsung phones `S5230` and `B5310` do not properly validate such multipart sequences. This allows us to craft messages that show up as a very large SMS message on the phone. When opening such a message the phone tries to reassemble the message and crashes. Depending on the exact model one to four SMS messages are needed to trigger the bug.

Micromax: The `Micromax X114` is prone to a similar issue like the Samsung phones but behaves slightly differently. When sending one SMS that contains a multipart UDH with a higher chunk ID than the overall number of chunks and a reference ID that has not been used yet, the phone receives the SMS message without instantly crashing. However a few seconds after the receipt the display turns black for some seconds before the phone disconnects and reconnects to the network.

4.4.9 Validation and Extended Testing

After the initial fuzz-testing we needed to validate our results over a real operator network since we tested in a closed environment – our own GSM network. We need to evaluate if the bugs can be triggered in the real world or if operator restrictions prevent this. For the validation we put an active SIM card (of the four German operators) into our test phones and connected them to a real mobile phone network. We sent the SMS PDUs that triggered the bugs using the AT command interface of another mobile phone. These tests validated all the bugs described in the previous section.

During our fuzzing tests we deactivated the security PIN on the SIM cards we used in the target phones so that we did not have to enter the PIN on every reboot. We also tested the phones with an enabled SIM PIN. Our goal was to determine if such reboots also reset the baseband and the SIM card. If the SIM card is blocked after reboot the phone is not reconnected to the GSM network, and, thus, the user is cut off permanently.

We determined that this is true for our LG, Samsung, and Nokia devices.

4.4.10 Bug Characterization

We group the discovered bugs depending on the software layer they trigger.

The first group are bugs that *require user interaction* such as the bug we discovered in the Samsung mobile phones. In this case the user has to view the message in order to trigger the bug.

The second group are bugs that crash *without user interaction*. These bugs occur as soon as the phone has completed receiving the entire message and starts processing it. In this group we put the bugs we found on the Motorola, LG, and Micromax devices.

The third and last group are bugs that trigger at a lower layer of the software stack. With lower layer we mean during the process of receiving the SMS message from the network. A crash *during the transfer process* means that the process is not completed and the network believes the message is not successfully delivered to the phone. We categorize the bugs discovered in our Nokia S40 and the Sony Ericsson devices in this third group.

4.5 Implementing the Attack

The attacks presented in this work utilize SMS messages to trigger software bugs and crash mobile handsets, interrupting mobile communications. These bugs cover the mobile phone platforms of all major handset manufacturers and a wide variety of different models and firmware versions. The resulting bug arsenal can potentially be abused to carry out a large scale attack.

4.5.1 Building a Hit-List

To launch an attack phone numbers of mobile phones need to be acquired since simply sending SMS messages to every possible number is problematic. Furthermore, sending SMS messages to a large number of unconnected phone numbers *dark address space* could trigger some kind of fraud prevention system, such as observed on the Internet to detect worms [80]. In addition, for the described attack only phone numbers that are connected to a mobile phone are of interest. Depending on the kind of attack, a

4.5 Implementing the Attack

different set of phone numbers is required. In one case an attack might be targeted towards a specific mobile operator, therefore, only phone numbers that are connected to the specific operator are of interest.

Regulatory Databases: In many countries around the world mobile network operators have their own area codes. Some examples are Germany[1], Italy[2], the United Kingdom[3], and Australia[4]. Such area codes can be readily acquired to help building a hit-list. Likewise one can use the *North American Numbering Plan* (NANP) to determine which area exchange codes are used by mobile operators.

Web Scraping: Web Scraping is a technique to collect data from the World Wide Web through automated querying of search engines using scripted tools. Finding German mobile phone numbers can be easily done through queries like `"+49151*" site:.de`. Moreover, online phonebooks [23] also include mobile phone numbers. These sites often allow wild card searches, and, thus can be abused to harvest mobile phone numbers.

HLR Queries: Some Bulk SMS operators [71] offer a service to query the Home Location Register (HLR) for a mobile phone number. These queries are very cheap (we found one for only 0.006 Euro) and answers the question if a mobile phone number exists and where it is connected. Together with the information from the regulatory databases one can easily generate a list of a few thousand mobile phone numbers that belong to a specific mobile network operator.

4.5.2 Sending SMS Messages

SMS messages can be sent by a mobile phone that provides either an API that allows it to send arbitrary binary messages or through its AT command interface. We used the AT interface for most of our testing and validation. To carry out any kind of large scale attack a way for delivering large quantities of SMS messages for low price is needed. Multiple options exist to achieve this:

---

[1] http://en.wikipedia.org/wiki/Telephone_numbers_in_Germany
[2] http://en.wikipedia.org/wiki/Telephone_numbers_in_Italy
[3] http://en.wikipedia.org/wiki/Telephone_numbers_in_the_United_Kingdom
[4] http://en.wikipedia.org/wiki/Telephone_numbers_in_Australia

4 Security Analysis of SMS Implementations on Feature Phones

Bulk SMS Operators: Bulk SMS operators such as [18, 39, 71] offer mass SMS sending over the Internet providing various methods ranging from HTTP to FTP and the specialized SMPP (Short Messaging Peer Protocol). Bulk SMS operators are so-called External Short Message Entity (EMSE) that are often connected via Internet to the mobile operators but sometimes have their own SS7 connection to the Public Switched Telephone Network (PSTN). Figure 2.1 shows the various connections of an EMSE. All Bulk SMS operators operate in the same way. For a given amount of money they deliver SMS messages to the specified destination(s). No questions asked. Most of the APIs support sending a single message to a list of recipients. Prices range from 0.1 to 0.01 Euro depending on the volume and destination of the messages. The APIs among the bulk SMS operators differ. Usually they allow to set a number of SMS fields from which they assemble the actual payload. Not all of them are offering the same predefined fields. For example [39] was the only one that allows us to set a TP-Protocol-Identifier field. However, we verified that the provided APIs are sufficient to carry out the presented attacks and to generate attack payloads that are identical to those sent from one of our phones.

Mobile Phone Botnets: A botnet consisting of hijacked mobile or smartphones [69] could also be used for such attacks since every mobile phone is capable of sending SMS messages. A mobile botnet has the distinct advantage of free message delivery and high anonymity for the attacker. using a mobile phone botnet one could circumvent restrictions Bulk SMS operator might have in different countries.

SS7 Access: With direct access to the Signaling System 7 (SS7) of the Public Switched Telephone Network (PSTN) an attacker can very easily send SMS messages in large quantities, for example to send SMS spam [16]. Figure 2.1 shows the basic network connections of a mobile network operator. SMS sending via SS7 also has the advantage of not being easily traceable, thus an attacker can stay hidden for a longer period of time. Additionally, SMS messages sent via SS7 are not restricted by the Bulk SMS Operators (APIs) in terms of content or header information that they contain.

4.5.3 Reducing the Number of Messages

There is one issue left with our attack. That is how can one determine the type of mobile phone that is connected to a specific phone number. If money does not play a role in

4.5 Implementing the Attack

carrying out the attack this issue is easily resolved. The attacker just sends multiple SMS messages, each one containing the payload for a specific type of phone, to each phone number. One of the messages will trigger the bug if the phone is vulnerable at all. This works well but is not optimal. To reduce the number of messages an attacker has to send we developed a technique that allows the attacker to determine what kind of phone is connected to a specific phone number. Actually we can only determine if a specific malicious message has an effect on the phone that is connected to a specific number.

Our method abuses a specific feature present in the SMS standard. This feature is called recipient notification, it is indicated through the TP-Status-Report-Request flag in an SMS message. If the flag is set the SMSC notifies the sender of the message when the recipient has received the message. Most Bulk SMS operators support this feature through their APIs. Our method works by measuring the delay between sending the message and receiving the reception notification.

The technique works as follows: First, we send the message containing the payload for *crash(1)*. Second, when we receive the receipt for that message we send the payload for *crash(2)*. Third, we measure the time difference between the two notifications. If the difference is equal we continue with the next payload. If the difference between both notifications is significant we determine that the first message crashed the phone. The phone needed to reboot and register on the network before being able to accept the next message. If there is no notification we determine that the phone did not receive the message because it crashed before completely accepting the message. Fourth, we continue until all crash payloads are sent. If none of them trigger, the phone number is removed from the hit-list. The method can be optimized through ordering the crash payloads according to the popularity of mobile phones in the targeted country.

With this method an attacker can optimize a hit-list during an ongoing attack by matching bug-to-phone-number. This optimized hit-list could as well be used for highly targeted attacks. For example against the network operator as described in Section 4.5.5, which explains our attack scenarios.

4.5.4 Network Assisted Attack Amplification

Some of the bugs we discovered prevent the phone from acknowledging the SMS message to the network. Figure 4.2 shows the states that happen during a message transfer from the network to the phone. In the case of some of our bugs (Nokia S40 and Sony

4 Security Analysis of SMS Implementations on Feature Phones

Ericsson; Bug Characterization Section 4.4.10) the message RP-ACK is not sent by the phone. This leads the network to believe that the message was not received, therefore, the SMSC will try to resend the SMS message to the phone. This re-delivery attempt is a perfect attack amplifier somewhat similar to smurf attacks [17] on IP networks.

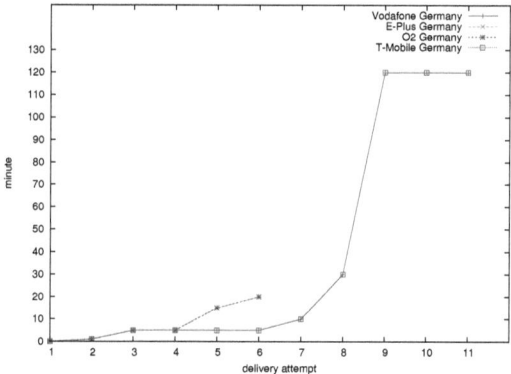

Figure 4.6: Timing of SMS message delivery attempts

In our tests, sending malicious SMS messages over real operator networks, we discovered that operators have different re-transmit timings, shown in Figure 4.6. Furthermore, they also seem to have different transmit queues. We measured the delivery timings of some German mobile network operators in order to determine how one could abuse the delivery attempts for improving our Denial-of-Service attacks. We conducted the test by attacking one of our Sony Ericsson devices and monitoring the phone using the Bluetooth method described in Section 4.4.4.

The tests were carried out on the networks of Vodafone, T-Mobile, O2 (Telefonica), and E-Plus. The initial delivery attempt is at minute 0. It shows that all operators do a first re-transmit after 1 minute, and a few more re-transmits every 5 minutes. In addition to what Figure 4.6 shows, Vodafone does an additional re-delivery 24 hours after the last delivery shown in the graph. O2 also attempts an additional re-delivery 20 hours after the last delivery shown in the graph.

Through the same test we determined that SMS messages are not queued, but have an individual re-transmit timer. That means an attacker can send multiple malicious SMS messages to a victim's phone with a short delay between each message and thus can increase the effect of the network assisted attack by sending multiple messages.

4.5.5 Attack Scenarios and Impact

There are multiple possible attack scenarios such as organized crime going after the end-user, the mobile operator, and the manufacturer to demand money. Attacks could also be carried out for fun by script kiddies and the like. Below we discuss some possible scenarios. We acknowledge that some scenarios such as the attack against individuals are more likely then an attack against a manufacturer.

Individuals: Individuals could be pressured to pay a few Euros in order to keep their phone operational. This has happened with the Ikke.A [69] worm that requested the user to pay 5 Euros in order to get back the control over their iPhone. In our case the victim could be forced to send a text message to a premium rate number in order to be taken off the hit-list.

Another attack against an individual or a group could aim to prevent them from communicating. This can be efficiently carried out if the target uses a SIM card with security PIN enabled, as we describe in Section 4.4.9.

Operators: Operators could be threatened to have all their customers attacked. Such an attack would mainly kill the operator's reputation as being reliable. The operator might also lose money due to people being unable to call and send text messages. In order to have a global impact such an attack has to be carried out on a very large scale for a longer time. As a result, customers could possibly terminate their contract with the operator. Such extortion scams were and still are popular on the Internet [21].

Furthermore, the operator's mobile network can be attacked directly or as a side effect of an large attack against its users. This could work when thousands of attacked phones drop off the network and try to re-connect at the same time. This can cause an overload of the back-end infrastructure such as the HLR. This kind of attack seems likely since mobile networks are not optimized for these specific kinds of requests. A similar attack based on unusual requests was shown in [85]. It is not normal that thousands of phones try to connect and authenticate at the same time over and over again. To optimize this DoS attack, the attacker needs to make sure to target phones connected to different BTSs and MSCs (Figure 2.1) of the targeted operator in order to circumvent bottlenecks such as the air interface at the BTS. A clogged air interface would throttle the attack.

Manufacturers: Likewise manufacturers could be threatened to have their brand name destroyed or weakened by attacking random people owning their specific brand of mobile phones. The attack could cost them twice. Once for the bad reputation and second for replacement devices. Even if the phones are not broken victims of such an attack will still try to claim their device broken to get a replacement.

Public Distress: A carefully placed attack during a time of public distress could lead to large scale problems and possibly a panic. One example occurred in Estonia [13] in 2007 when a group of people carried out a Denial-of-Service attack against the countries Internet infrastructure. Additionally, cutting off certain user groups such as fireman or police officers during an emergency situation would have a critical impact. Not every country has special infrastructure for emergency personal, and, therefore, rely on mobile phones to communicate. This is even true in countries like Germany where every police officer carries a mobile phone since their two-way-radios are often not usable.

4.6 Countermeasures

In this section we present countermeasures to detect and prevent the kind of attacks we developed. First, we present a mechanism to detect our and similar attacks through monitoring for a specific misbehavior. Second, we discuss filtering of SMS messages. Filtering can be done on either the phones themselves or on the network. We discuss the advantages and disadvantages of each of them. Third, we briefly discuss amplification attacks.

4.6.1 Detection

To prevent our attacks, operators first need to be able to detect them. Detection is not very easy since the operator does not get to look inside the phone during runtime. Therefore, the only possible way to monitor the phone is through the network. We propose the following:

Monitor Phone Connectivity Status: Monitor if a phone disconnects from the network right after receiving an SMS message.

Log last N SMS Messages: Log the last N SMS messages sent to a particular phone in order to analyze possible malicious messages after a crash was detected. Use the message as input for SMS filters/firewall.

Use IMEI to Detect Phone Type: The brand and type of a mobile phone can be derived from the IMEI (International Manufacturer Equipment Identity). This is useful to correlated malicious SMS messages to a specific brand and type of phone.

Using this technique it is possible to catch malicious SMS messages that cause phones to reboot and lose network connectivity. *This should especially help to catch unknown payloads that cause crashes.* Such a monitor is also capable of detecting if a large attack is in progress by correlating multiple SMS-receive-disconnect events in a certain time-frame.

4.6.2 SMS Filtering

SMS filtering can be implemented either directly on the phone or within the operator's network. Both possibilities have inherent benefits and drawbacks that are presented in this section.

It is important to reconsider the process of SMS delivery. First, an SMS message is sent from the sender phone to the senders SMSC. Next, the senders SMSC queries for the SMSC of the recipient and delivers the message to the responsible SMSC. Finally, the relevant SMSC locates the recipient's phone and delivers the SMS message via the BTS Over-the-Air.

Client-side SMS Filtering would need to be done right after the modem of the phone received and demodulated all the frames carrying the SMS message and before pushing it up the application stack. The filter would need to parse the SMS message and check for known bad messages similar to signature-based antivirus software or a packet filter firewalls. The problem with this solution is the update of the signatures. Of course, the parser in the SMS filter must be bug free otherwise the attack just moves from the phone software to the filter software. Also, devices that are already in the field would not profit from such a filter since only new phones will have this. Also, newer phones will likely not contain bugs that are known at the time they are manufactured. Therefore, we believe network-side filters make more sense.

Network-side SMS Filtering takes place on the SMSC of the mobile network operator. Therefore, it can inspect all incoming and outgoing SMS messages. There are multiple advantages of network-side filtering. First, the filter software runs on the network, therefore, it covers all mobile phones connected to that network. Second, changing the filter rules can be done in one central place. Third, malicious SMS messages are not sent out to the destination mobile phones, therefore, reducing network load during an attack.

Network-side filters also have drawbacks. First, if a phone is roaming within another operator's network, the SMS message does not travel through the network of the home operator. Thus the filters are not touched. This is the only advantage of phone-side SMS filtering. In this case the user becomes attackable as soon as he leaves his home network. For traveling business people in Europe, this is quite normal. The GSMA already has a solution for this issue called SMS homerouting. *SMS Homerouting* as specified in [7] defines that SMS messages are always routed through the receiver's home-network. Meaning that all SMS messages travel through SMSCs of his service provider at home. SMS messages, therefore, can be filtered by the receiver's service provider. The second issue with network-side filtering is privacy. In order to do SMS filtering the operator must be allowed to inspect SMS messages. This could be an issue in some countries where mobile telephony falls under special regulations.

4.6.3 Preventing Network Amplification

Attack amplification through re-transmissions of SMS messages should be avoided since this greatly helps an attacker. We suggest that operators limit the number of re-transmissions. Some operators re-send the messages 10 times, this seems unnecessary.

4.7 Conclusions

In this work we have shown how to conduct vulnerability analysis of feature phones. Feature phones are not open in any way, the hardware and software are both closed and thus do not support any classical debugging methods. Throughout our work we

4.7 Conclusions

have created analysis tools based on a small GSM base station. We use the base station to send SMS payloads to our test phones and to monitor their behavior. Through this testing we were able to identify vulnerabilities in mobile phones built by six major manufacturers. The discovered vulnerabilities can be abused for Denial-of-Service attacks. Our attacks are significant because of the popularity of the affected models – an attacker could potentially interrupt mobile communication on a large scale. Our further analysis of the mobile phone network infrastructure revealed that networks configured in a certain way can be used to amplify our attack. In addition, our attack can be used to not only attack the mobile handsets, but through their misbehavior can be used to carry out an attack against the core of the mobile phone network.

To detect and prevent these kind of attacks we suggest a set of countermeasures. We conceived a method to detect our and similar attacks by monitoring for a specific behavior.

4 Security Analysis of SMS Implementations on Feature Phones

5 Cellular Malware Communication Capabilities

In this chapter we discuss the security impact of malware that has unlimited access to the cellular modem. We create a Proof-of-Concept cellular botnet that facilitates the cellular modem for command and control.

5.1 Introduction

It is clear that mobile and smartphones are the future of personal computing — just as the personal computer was many years ago, and additionally also their IP connectivity follows this trend, i.e., more and more phones have a pervasive IP connectivity, i.e., WiFi, GPRS, EDGE, 3G, etc.

Sadly enough, this success of an (almost) single platform architecture (mainly MS Windows-based) installed on a nearly infinite number of IP-connected PC's has led to an unforeseen security crisis for the PC platform which culminated in one of the largest security threats for the IP world: large scale criminal botnets [85].

Given this similarity between the PC platform and the emerging (and dominating) smartphone platforms like iPhone, Android (Google) phone, and Windows Mobile, it is a legitimate question whether the cellular world could also enter a severe security crisis like the PC itself? Especially, we are interested in answering this question with regard to the existence of practical and functional cellular botnets.

The practical existence question is especially important, as the theoretical threat of cellular botnets was just recently investigated and emphasized by Traynor et al. [85] — simply assuming the theoretical existence of such botnets. Moreover, their focus was on the security impact for the fragile and complicated cellular architecture on which we are all depending on — day by day. Their main research result showed that a relatively small number of cellular bots can already force the collapse of a targeted victim core network. Interestingly, they smoothly concluded that the challenges for a functional and large-scale cellular botnet are noteworthy and that such botnets might not be too quickly seen in the wild.

Thus, the present work perfectly complements their research from the cellular platform side, as we solve their cellular challenges and describe the architecture and even the implementation of a practical and simply realizable cellular botnet for the iPhone.

Especially, we show how we designed, implemented and evaluated an iPhone-based mobile botnet. We did this to understand what it takes to build a botnet that resides on mobile phones and on a mobile phone network. We think this is an important first step in order to start thinking about urgently needed counter measures for mobile phone botnets.

We started by following the current developments in botnet research and built a Peer-to-Peer (P2P) based mobile phone bot. The P2P bot was quite simple to design and implement, and, therefore, presents an easy path for an unskilled botmaster.

Diving deeper in to the specifics of mobile phone botnet we further created a Short Message Service (SMS) [6] based bot. A bot that can be controlled entirely via SMS. We further improved our SMS bot by turning it into a hybrid of SMS and HTTP in order to reduce the number of SMS messages that need to be sent for controlling the bots.

In the end we showed how powerful a mobile phone botnet could be if one combines the P2P with the SMS-HTTP hybrid approach. A bit frightened by the success of our cellular bots we recognized that this hybrid bot would be very hard to be detected and stopped if controlled by a skilled botmaster. Thus, we also stopped our further research at this point as our main questions and motivations were completely solved.

This chapter makes the following main contributions:

- We showed a cellular botnet architecture and even evaluated it with several practical implementations.

- We solved the environmental challenges of such cellular botnets.

- We implemented and evaluated a P2P-based command and control mechanism for mobile phone botnets. Our bot implements the Kademlia P2P protocol and joins the Overnet network.

- We designed, implemented, and evaluated multiple SMS-based C&C mechanisms. The SMS approach raises the bar for the anti-botnet community.

- We created communication strategies for mobile phone-based botnets. The strategies are designed to increase the stealthiness of mobile phone botnets.

Chapter Organization

The rest of this chapter is structured in the following way. In Section 5.2 we show how easy it can be to hijack many thousand iPhones using the Internet. Section 5.3 discusses the intrinsic challenges that cellular networks pose for botnets. In Section 5.4 we present our command and control mechanisms for mobile botnets, while Section 5.5 continues elaborating on our communication strategies for mobile phone botnets. Eventually, Section 5.6 details our Proof-of-Concept implementation of our mobile bots including a self-critical evaluation. Finally, Section 5.7 draws some important conclusions.

5.2 Howto hijack many thousand iPhones

In November 2009 somebody exploited the facts that jailbroken[1] iPhones get a default root password assigned, often have the secure shell daemon (sshd) installed, and get an public IP address assigned to create a mobile phone worm. The worm was named ikee.A [69] and infected around 22,000 iPhones within two weeks by simply copying itself via secure copy (part of ssh) from iPhone to iPhone. Later somebody added a very simple command and control mechanism to ikee to turn it into a botnet, this botnet was called ikee.B. The command and control mechanism was simply polling a web server to download and run a shell script.

This example shows how easy it is to hijack many thousand mobile phones through the Internet without any special knowledge about mobile phones or mobile phone security. Therefore we believe that this was just a first taste of what will happen in the future. Also if you look at vulnerabilities like the one we presented in Chapter 3 through witch an iPhone could have been hijacked through SMS it becomes clear that mobile botnets are sure to come to existence.

---

[1] http://en.wikipedia.org/wiki/Jailbreak_(iPhone_OS)

5.3 Cellular Challenges

Mobile and smartphones present a number of challenges that need to be meet in order to design a botnet that is able to exist and thrive in the mobile phone environment. The problems range from: 1) limited run time due to the use of batteries as the power source, to 2) connectivity problems due to the absence of public IP addresses, 3) constant change of connectivity, 4) the problem of diversity of mobile phone platforms, and 5) the costs of mobile communication. In the following we will discuss these problems in further detail.

5.3.1 Absence of Public IP Addresses

Public IP addresses are needed for direct communication of bots. Without public IP addresses an intermediate communication hub is required, unfortunately most mobile phone service operators put their customers behind a Network Address Translation (NAT) gateway and thus the devices are not directly reachable. Although the attack vector presented in Section 5.2 shows the picture of a mobile operator providing public IP addresses to customer phones, this is not the common case. Even if a mobile operator chooses to provide public IPs to his customers, mobile phones will still sit behind a NAT gateway for many hours during the day. This is the time the user spends at home where his phone is connected to the local wireless LAN in order to benefit from higher Internet speeds and to lower the services charges by using his DSL or cable line.

5.3.2 Platform Diversity

The size of a mobile phone botnet will be relatively small compared with botnets based on hijacked desktop computers. The main reason for the size limitation is related to the diversity of mobile phone platforms, therefore we think each mobile phone botnet will be targeted towards a specific device, platform, or platform version. Due to the small number of bots in a mobile phone botnet it will be hard and maybe impossible to build an independent communication infrastructure such as P2P network that exclusively consists of hijacked mobile phones.

Connectivity	Hours
WiFi	Early morning (still at home)
GSM/3G	Morning (travel to work/school)
GSM/3G	Day time (while at work/school)
WiFi	Early evening (back at home)
GSM/3G	Early Night (going out)
WiFi	Night (bed time)

Table 5.1: Connectivity times.

5.3.3 Constant Change of Connectivity

Constant change of connectivity is something that is normal for a mobile phone compared with a desktop computer that is connected to the Internet via a DSL line. The connectivity of a mobile phone changes for many reasons. First, mobile phones move around the physical world. Their wireless connection comes and goes depending on the position of the device and the available type of mobile network capabilities. GPRS vs. EDGE vs. 3G. Individual phones might be disconnected for a relatively long time even though the phone itself is powered up. Second, is the earlier mentioned use of local wireless networks, this again would change the connectivity properties of a mobile bot. Therefore, a mobile phone botnet is likely to be very unstable in terms of the size and the kind of network connectivity of an individual node. Table 5.1 shows the connectivity times of the mobile phones of the authors and some of their colleagues.

5.3.4 Communication Costs

In the world of mobile telecommunication most types of communication result in costs for the ones who communicate. These costs have to been taken into account when designing a botnet Command and Control mechanism since a significant rise of the phone bill will lead to investigation of the cause and thus may lead to detecting the bot infection. Especially interesting is SMS, since here each message sent costs money. Also deepening on the type of mobile phone contract SMS messages can be completely free when sent to subscribers on the same network. Further things such as roaming has to be considered since the charges for communication are significant higher during roaming. Mobile-data usage might be disabled during roaming but services like SMS still work. A mobile phone bot therefore might need to query the roaming status in order to fit in with the other applications running on the device.

5.4 C&C for Mobile Botnets

Command and Control (C&C) is the most important part of a botnet. For the botmaster it is the path to deliver commands to his botnet and for the defender it is the major attack vector in order to dismantle and destroy a botnet. The C&C channel therefore has to be carefully designed to be reliable for command delivery as well as resilient against many kinds of attacks.

In [76] the authors use Bluetooth as the transport channel for Command and Control of their mobile phone botnet. We believe that a botnet based on local wireless communication will be large enough to be of any use for a botmaster, therefore, we focus our research on Internet and mobile phone network based C&C.

Over the past years there has been some major development in botnet C&C. Earlier botnets used IRC (Internet Relay Chat) for C&C but countermeasures against IRC-based C&C such as *Botnet Tracking* [32] has made IRC useless for C&C. The botnet can be tracked down by analyzing the bot to identify the IRC channels to finally find the master server that can be taken down to destroy the botnet. Today most botnets use some kind of P2P scheme for C&C such as discussed in [35]. P2P-based C&C botnets are more resilient against attacks then IRC-based botnets but still can be tracked using methods like the *Sybil* attack [24] and infiltrated as happened with the Storm Worm botnet [41]. Still we decided to implement a P2P-based approach since this is currently is the best known schema for IP-based C&C. Our goal here is to show that P2P-based C&C works for mobile botnets.

In our work we have followed two major paths for C&C. First, we evaluated a P2P-based approach since this seems to be the current overall trend in botnet research. Also we did not create our own P2P network as suggested in [42]. The second path we followed is a SMS-based approach. *We chose SMS because we think that SMS communication is much harder to observe, analyze and disrupt by security researchers and the anti-botnet community, and, therefore, it is likely to be chosen as the C&C channel by knowledgeable botnet creators.* We actually designed two SMS-based C&C mechanisms to get a broader overview of the possibilities and problems of SMS-based bot communication.

5.4 C&C for Mobile Botnets

In the following we will first discuss the P2P-based approach since it is a bit simpler than the approach based on SMS. Before we discus the actual C&C channel we briefly talk about the additional required features in order to secure the commands sent over the C&C channel.

5.4.1 Securing the C&C Communication

In order to protect the commands sent over the Command and Control channel from tempering all commands carry a digital signature using public-key cryptography. Further to prevent replay attacks, commands carry a sequence number. Only commands that carry a sequence number that is higher then the one from the last accepted command will be accepted as a valid command.

5.4.2 Peer-to-Peer C&C

For our P2P-based approach we choose Kademlia [55] as the protocol and Overnet[2] as the P2P-network to join. We chose to rather join an existing network instead of creating our own because of the mobile phone related problems and issues that we discussed in Section 5.3. The main reason being the unavailability of a stable set of public IP addresses.

The basic design idea for our P2P-based Command and Control channel is to use the P2P network as a kind of rendezvous point. The P2P network is only used as a basic communication channel using the publish and search functionality of the distributed hash table (DHT). The botmaster publishes a `command` to the P2P network and the bots search for a specific key in order to retrieve the `command`. The publish and search functionality is solely based on functionally offered through the DHT, and, therefore, no actual file sharing functionally needs to be present on either the botmaster nor the bot side. Figure 5.1 shows a high level view of a P2P-based mobile phone botnet.

Battery consumption plays a very important role in the mobile phone world, therefore it is very important for a mobile phone bot to not drain the battery significantly. A significant battery drainage will otherwise lead to detection of the bot rather easily. Battery drain is mostly related to two operations, high CPU load and heavy radio usage. Our main concern is the radio usage. In order to reduce the network activity of our bot it connects only briefly to the P2P network to search for the key that results in the

---
[2]http://en.wikipedia.org/wiki/Overnet

5 Cellular Malware Communication Capabilities

Figure 5.1: Kademlia P2P C&C

command from the botmaster. After the search, the bot quickly disconnects from the P2P network. Initially we designed the bot in a way that it connects to the Peer-to-Peer network about every 15 minutes. Upon connection it waits until the connection has stabilized and is ready to fire search queries (this seems to take between 30-60 seconds). In a 20 second interval in searches three times for the rendezvous key and then disconnects. This communication pattern is very similar to a background email poll, and, therefore, should not cut to deep into battery consumption. For times where the botmaster needs faster responds times for his botnet he can issue a command that changes the time interval of connecting to the P2P network. The interval can of course also be increased for less battery consumption and lower responds times (for times where the botnet is not heavily used).

5.4.3 SMS C&C

In this section we present our two SMS-based C&C mechanisms. Both schemas are based on the fact that the botmaster has a complete list of bots or actually a list of phone numbers that correspond to the bots, at all times.

Below we will first provide a brief introduction to the Short Message Service, then we will discuss each SMS C&C schema and finally we will briefly talk about obfuscation of the C&C SMS messages.

The Short Message Service is one of the basic services of the mobile phone network. SMS is used for text messaging by users and for background services that are not directly visible to the user. SMS supports transport of binary data, and, therefore, can be used to send arbitrary data such as Command and Control information for a botnet. Although SMS messages are limited to 140 octets each, we show that this is enough for a highly flexible and secure botnet communication. In this work we will not discuss the details of the Short Message Service itself and will stick to the parts that are important for the design of the C&C channel.

The basics of SMS communication are. The sender only needs to know the phone number of the receiver in order to send him a message. To send a message, the sender encodes the phone number together with some flags and the payload in to the SMS PDU format and hands it over to the mobile phone modem using the GSM AT command set. The modem takes care about delivering the message to the mobile phone network.

In the network SMS messages are handled by the Short Message Service Center (SMSC). The SMSC forwards the message to the receiving mobile phone. If the receiving mobile phone is switched off the SMSC buffers the message until the receiver is switched back on. The receiver, upon reception of the SMS, extracts the payload from the PDU. The payload is just a number of octets at the end of the PDU.

SMS-only C&C

In this scenario all communication from the botmaster to the botnet is carried out over SMS. There are a few exceptions such as an update of the bot software or data transfer back to the botmaster which are still carried out over IP. Sending SMS messages costs money in most cases (see our discussion on SMS in Section 5.3.4), therefore it does not make sense for the botmaster to send messages to each bot directly. In this section we describe our SMS-only communication schema. We separated the schema in four parts: infection, communication, repair, and management.

- *Infection* takes part in three steps. In the first step the bot-software is installed on the hijacked phone using a software or configuration vulnerability. In the second phase the newly installed bot sends an SMS message to its infector, the infector provides his own phone number during bot installation. The SMS is sent in order to determine the phone number of the new bot. Sending a SMS message is the only reliable way to determine the phone number of a mobile phone, since it is not necessarily stored on the SIM card or on the phone itself. In the third and last step the infector dumps his list of phone numbers of devices he infected to a drop-site to be collected by the botmaster.

5 Cellular Malware Communication Capabilities

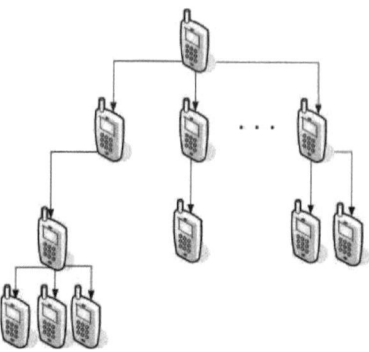

Figure 5.2: SMS only C&C

- *Communication* takes place in a tree model as shown in Figure 5.2. Meaning the botmaster sends a command message to the root node of his botnet. Each individual bot forwards the message to all bots known to them.

- *Repair* takes place after the botmaster determined that the communication tree has broken at some point. In order to determine if the tree is intact once in a while the botmaster sends a *broadcast ping* that every node needs to answer. Nodes that fail to answer the ping message are removed from the tree. If a none leave node is removed all its sub-nodes are reassigned to other nodes. This is done by sending a message containing a list phone number(s) to an active node. Since the botmaster has at any time a complete overview of his botnet he can carry out a more intelligent repair phase by checking smaller sub-trees instead of the whole tree (the whole botnet) at once.

- *Management* of the botnet is required since it must be taken care of that a single node does not have too many direct sub-nodes. Each direct sub-node will require one SMS message to be send to in case of a message being forwarded. Further if a node with many direct sub-nodes disappears all the direct sub-nodes need to be moved to another node, leading in more SMS messages being sent.

In the ideal case the botmaster only needs to send out one SMS message to reach every node in the botnet, also he might not even need to send the message himself but rather have a hijacked phone send the initial message.

5.4 C&C for Mobile Botnets

The SMS only design has a weak point, that is the existence of node lists (phone numbers) in most of the bot hosts. Therefore, making it easy for an attacker or anti-botnet researchers to warn the individual owner of an infected phone by simply sending him an SMS message. In order to partially prevent this from happening and to improve and ease the management and repair steps we designed a SMS-HTTP hybrid communication schema. This schema is discussed in the next Section.

SMS-HTTP Hybrid

After realizing that a SMS-only-based Command and Control channel bears certain problems and the fact that IP communication is still required to accomplish any meaningful data transfer we designed a SMS-HTTP hybrid C&C channel for our mobile phone botnet.

The SMS-HTTP hybrid design additionally improves the SMS-only design in following ways. First, it removes the necessity to keep information about the botnet at each of the nodes. Therefore, making it a bit more resilient against attacks. Second, it eases the botnet management and repair by moving these task from the botnet to the botmaster. Third, it splits up the botnet in to multiple subnets and thus makes it harder to be detected.

The hybrid schema shares many properties of the SMS-only schema. The infection part of the hybrid schema works in exactly the same way. The only difference is that both, the newly infected bot and the infector, do not store the phone number of each other. The newly infected bot deletes the phone number of its infector after sending him the SMS message to determine his own phone number. The infector deletes the phone number of the newly infected phone right after dumping it at the drop-site in order to be collected by the botmaster.

The basic idea of the hybrid schema is to split the communication in to a HTTP and an SMS part. Command SMS messages are pre-crafted by the botmaster and are uploaded as encrypted files to some website. The URL to these files are then sent to random bots of the botnet via SMS. The bots download and decrypt the files and sent out the pre-crafted SMS messages. The encryption key is part of SMS message that contains the URL to the file. Figure 5.3 shows the three steps of the SMS-HTTP hybrid communication. In a decent sized botnet the first round of pre-crafted SMS messages could again contain URLs to another batch of pre-crafted command messages. Due to the fact that in the SMS-HTTP hybrid there is no fixed structure through that all

5 Cellular Malware Communication Capabilities

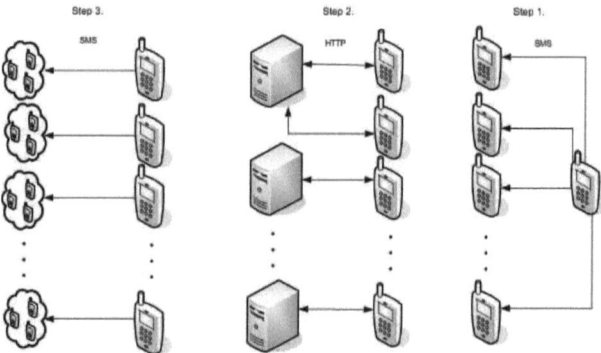

Figure 5.3: SMS-HTTP hybrid C&C

communication is happening it is quite hard to determine if a botnet is active on a mobile phone network by just looking at the SMS traffic.

The repair part of the hybrid schema works in the same way as the repair part of the SMS-only schema. The botmaster has to regularly probe each bot to determine if it is still part of the botnet. The difference is that the botmaster does not need to move individual bots around the botnet in order to keep them in the working communication tree. Hosts that are no longer part of the botnet are just removed from the global bot list and thus are not considered the next time a command is sent out to the botnet. The management part as such does not exist in the hybrid design.

Obfuscation

One idea for obfuscation is to encrypt all C&C SMS messages with a symmetric key. The key would be globally known by all bots, and, therefore, it will not prevent reverse engineering or off-line command analysis. But it will make life much harder for the mobile operators to filter out the messages, since they can not really tell what kind of message they see. To prevent hard coding of the key in to any IDS and anti-virus system there will be regular key updates. The key updates must be frequent enough so that C&C message parsing is required in the IDS. This will make it more costly to keep track of the botnet.

5.5 Communication Strategies

Communication is the most important part of a botnet, especially on a mobile phone since mobile phones have limited resources. A battery that drains faster than it used too, a high phone bill, slow or clogged 3G data can easily lead to detection and removal of the bot. In this Section we discuss our ideas for how a mobile phone botnet should carry out it's communication in order to stay hidden and still have maximum functionality.

5.5.1 IP-based Communication

Even with a SMS-based C&C channel a bot still requires IP-based communication in order to transport larger chunks of data to and from the hijacked device. The data can be anything ranging from harvested information to a software update of the bot itself.

The problem with bulk data transfer on mobile phones is that the mobile connection can be slow, and, therefore, the transfer will take time and thus becomes detectable by the user. Especially if the user is also trying to use the network connection at the same time. If done regularly the costs might show up on the phone bill.

Internet Peer-to-Peer based botnets will more or less constantly communicate using IP packets, therefore, in order to decrease the possibility for detection the IP-based communication should be kept as hidden as possible.

We developed some strategies for IP-based communication for mobile botnets in order to keep the bots as hidden as possible. The main idea is to communicate mainly using the mobile phone network since this is somewhat harder to monitor. Bulk data should only be transfered using WiFi, if possible.

First, a bot should only initiate a bulk data transfer when connected through WiFi or a high speed 3G network. The 3G network should only be used if the bot some how determined that the phone it is running on has not used any WiFi network for some amount of time. This might be necessary since some mobile phone users will not pay for mobile-data usage, and, therefore, will not bother to use WiFi at all.

Second, all background communication (such as P2P chatter) should be carried out over the 3G network. The P2P chatter does not produce any significant amount of traffic. Also file transfer is not happening at all. This is in order to avoid detection of P2P traffic on the WiFi and DSL link. Also we anticipate that mobile phone network operators do not really monitor the traffic on their network. At least not traffic that does not require a lot of bandwidth, such as the P2P chatter.

5 Cellular Malware Communication Capabilities

Third, all bulk data transfer should be carried out using HTTP. This is to avoid blocked ports and any kind network monitoring.

5.5.2 SMS-based Communication

Each SMS sent might produce costs on the sender and on the receiver side, depending on the contract. Therefore, we created a set of rules in order to reduce the number of SMS messages to send. Also not only the number of messages sent need to be considered but also the destination of the message. Destinations such as foreign countries are likely to be more expensive, but also phone numbers that are routed on another operators network could introduce more costs.

In order to design a useful strategy for sending SMS messages one has to analyze the most common mobile phone contracts that are attached to the targeted mobile platform. For example a big German mobile operator offers four different contracts for the iPhone. Only the contract with lowest monthly rate charges for individual SMS messages. The other contracts include free SMS messages sent within the same network. The contract with the highest monthly charges include 3000 SMS messages sent to any destination.

Taking these facts into account we came up with a simple rule for SMS communication. *Grouping of bots by country and by operator. Limit sending SMS messages between these groups to a minimum.* Messages sent within an operator can be considered free.

Determination of country is easy because of the country code. Determination of the operator is a bit more complex. In certain countries this can be done by looking at the mobile number area code. If it is not possible to determine the operator from the area code each bot can still query the SIM card for the operator name.

This information needs to be communicated back to the botmaster. This could be done during infection time since here the infector has to deliver the phone number of the new bot to the botmaster anyways.

5.5.3 Data Delivery

Mobile phone resident bots have access to interesting data. In order to transport data from the device back to the botmaster the bot encrypts the data to avoid detection during either transport or through a raid on the drop-site. The encryption is done using public key cryptography to prevent data decryption by extracting the key from a bot infected phone. Therefore, each bot carries the botmaster's public key and uses it to encrypt a random symmetric key that is used for data encryption.

5.6 Bot Implementation

We created a Proof-of-Concept implementation of our bot design. The implementation includes both the SMS and the P2P-based Command and Control schemas.

We begin with a general description of how command-packets are structured in our botnet. The packet-format is designed in a way that fits both the P2P and SMS-based approaches. We chose to do this since our overall goal was to build a super-hybrid bot that features both C&C schemas. This way the botnet becomes more flexible and very hard to disrupt. The actual implementations are still separated but with spending a little more time on the implementation the super-hybrid bot can be easily build.

The two bots basically are composed of a single executable, a ECC public key to be used for command authentication using ECDSA, and a RSA public key for encrypting data sent from the bots to the botmaster. The P2P version additionally carries a initial peer-list for the Overnet P2P network. The SMS-client carries an additional dynamic link library for library injection.

5.6.1 Commands

Commands are composed out of four elements, shown in Figure 5.4. The command type, the command itself, a sequence number, and a signature. The command type simply specifies what kind of command the packet contains, this can be a shell sequence such as `ping -c 3 www.wired.com`. We will discuss some of the important command types later. The sequence number is a 32-bit counter to prevent replay attacks of commands. The bot will only execute commands with a sequence number higher then the one he has stored. The command packeted is signed and the signature is stored in the packet. The signature guarantees that only the botmaster can send commands to the botnet. In order to keep the command packets as small as possible ECDSA [47] is used for signing. ECDSA signatures are between 70 and 72 bytes, and, therefore, fit in to SMS messages while still leaving space for the actual command. In order to be able to use ECDSA on the iPhone we had to build our own versions of libssl and libcrypto, these were statically linked to our bot executable.

Content	Bytes
Signature length	1
ECDSA Signature	variable
Sequence Number	4
Command Type	1
Command	variable

Figure 5.4: Basic command structure

5.6.2 Kademlia P2P Client

We based our Peer-to-Peer bot on the KadC[3] Kademlia implementation. We chose KadC because it has no dependencies other than a minimal POSIX API which makes it highly portable. In theory one should be able to compile it for all current smartphone operating systems (including Android, Windows Mobile and Symbian). Further KadC only implements the DHT and not the file transfer part of the P2P network, thus making it the perfect candidate for our purpose. Below we discuss how we use Kademlia and Overnet as our C&C channel.

Once our bot joins the network it starts searching for a specific hash every 15 minutes. In order to send a command to the botnet the botmaster publish a entry in the DHT using the hash the clients search for. The command is transported using the meta information that can be published together with the hash. If the hash is found the client extracts the command from the meta information. The command-data is stored inside the meta information returned with the search result. Since Kademlia only seems to support ASCII data in meta information the command-data is based64 encoded. Although we had to change the actual alphabet used for encoding since Kademlia does not support uppercase characters.

5.6.3 SMS Client

We implemented the SMS bot-client to piggy back on the iPhone's telephony stack (`com.apple.CommCenter`). This was done using library pre-loading to sit between the iPhone's modem and the telephony stack in order to intercept SMS messages before the SMS application sees them. We use the technique that is described Chapter 3.

---

[3]http://kadc.sourceforge.net/

5.6 Bot Implementation

The technique also works on other smartphone platforms such as Android and Windows Mobile, therefore we think this is a reasonable approach. The pre-loaded library monitors open(2) calls and replaces the file descriptors for the modem lines with file descriptors connected to the actual bot application. Thus all AT commands and results to and from the modem first pass through the bot. If the bot recognizes an incoming SMS message it tries to parse it, if the parsing is successful the SMS is so to say swallowed and not passed on to the telephony stack. All other SMS messages are passed on. Therefore, the control SMS messages never touch the SMS application or SMS database and stays hidden.

Sending SMS messages is done by issuing AT commands to the modem device (/dev/tty.debug). This again is not seen by the user in anyway since the SMS message is not handled by the telephony stack.

5.6.4 Evaluation

We evaluated our bot design and implementation by installing the bot on a number of iPhones in our lab. The bot did not have any kind of spreading functionality implemented in order to make sure it does not escape our test environment. Also the evaluation was focused on the Command and Control (C&C) functionality, rather than the infection routines.

The evaluation was conducted by running the bot and sending it commands, either via the P2P network or directly via SMS. We constantly monitored the bot activity in order to determine if the commands were successfully received and executed. We did not only sent correct commands to the bots but also commands with broken signatures and invalid sequence numbers.

For evaluation we implemented a number of commands, these are:

- *Add phone number(s)*. This command adds a list phone numbers to the forwarding list of a bot.

- *Set sleep interval*. This command is used to set the sleep time between connecting to the P2P network for searching for commands.

- *Execute shell sequence*. This command is used to execute a shell sequence.

- *Download URL.* This command is used for the SMS-HTTP hybrid to download a command file. Besides the URL the command also includes a 128-bit key which is used to decrypt the downloaded file.

Below we will first discuss the P2P-based approach followed by the SMS and SMS-HTTP hybrid design.

Kademlia P2P

We evaluated two scenarios, one where the devices are connected to the Internet via WiFi and one where the devices are connected using a mobile-data connection. The botnet was controlled by a special version of the bot that can issue commands, this version was running on a laptop connected to the university network.

We ran several tests where we executed shell commands and changed the sleep interval for connecting to the P2P network. A basic test was to `ping` one of our servers on the Internet, here we could easily monitor that all our bots actually executed the command.

All in all we where more then satisfied with the performance of our P2P bot-client, especially since it was rather easy to implement.

SMS and SMS-HTTP Hybrid

In order to evaluate and test the SMS-based C&C design we implemented a small tool that crafts a command SMS message for our botnet. The tool takes the phone number, the type of command, and the command parameters as input and generates a ready to send SMS PDU. The PDU can then either be sent via the GSM AT command-set or be stored in a file to be used for the SMS-HTTP hybrid C&C mechanism.

We ran a number of tests in order to verify that our SMS C&C mechanism actually works. We again ran the ping-based test to verify that commands with a correct signature and sequence number are accepted. We further verified the basic functionality of file downloads by submitting a URL download command. Forwarding of command messages also worked as expected.

5.7 Conclusions

Through our work we practically confirmed that the theoretical threat of mobile botnets as pointed out by Traynor et al. [85] is real and concrete. We determined that it is easily possible to create a fully functional mobile phone botnet on the most popular smartphone — Apple's iPhone. A mobile phone botnet has many similarities with a desktop computer based botnet but also has certain properties that need to be considered in order to keep the botnet running and hidden. We investigated those cellular specific challenges and properties. We determined that the hybrid approach of SMS and HTTP is the most promising and most dangerous botnet Command and Control structure. Our successful mobile bot implementation stresses that this mobile specific hybrid approach would require very specific and difficult counter measures from a telco. The reason is that two totally different cellular subsystems, i.e., SMS and IP, would be needed to monitor and synchronized for specific, but yet unknown events and messages. This would cause cumbersome burdens for a telco to detect and prevent such mobile botnets. Given our preliminary but devastating results from our research journey we feel that there is an urgent need for novel and appropriate cellular phone and network protection mechanisms.

5 Cellular Malware Communication Capabilities

6 Improving the Security of the Cellular Modem Interface

In the preceding chapters we analyzed the security properties of the cellular modem interface and developed Proof-of-Concept attacks that leverage it. In this chapter we present a system to control access to the cellular modem. Our respective architecture prevents malware from carrying out attacks by abusing parts of the cellular modem interface.

6.1 Introduction

In the past years a lot of effort has gone into securing smartphones. There are academic contributions [25, 64, 94] and work performed by smartphone operating system (OS) vendors such as Apple, Google, Symbian, RIM or Microsoft. However, the efforts concentrated on the OS, to protect users from attacks and to mitigate malware such as Trojans.

Despite recent attacks, which target the cellular core network, few methods of defense are known. These attacks are based on hijacked mobile phones (mobile botnets) that produce signaling traffic sent from mobile phones to the cellular network core conducting Denial-of-Service attacks. These attacks demonstrate that current security improvements seek to protect the actual device and not the environment in which they operate, namely, the cellular core network.

Related security and reliability problems are caused by rooted (fully user controlled) smartphones. The problem is that rooting disables protection mechanisms of the OS, allowing the user to install arbitrary applications to his device. Such applications might leverage extended access privileges and may use them for intentional malicious activity and accidental harmful operations.

In this work we present our novel solution for protecting the cellular network infrastructure from malicious smartphones. Our protection system is called the *virtual modem*. It secures the baseband (the cellular modem), the entity that communicates with the cellular network. To the best of our knowledge nobody has yet attempted this path for securing cellular communication.

In contrast to a network side solution, our protection system is designed to run on the mobile phone. Changes to the cellular network equipment are very expensive and time consuming which would result in a slow adoption of any newly proposed protection mechanism. On the other hand, smartphone development cycles are very short. New smartphones are brought to the market every 6 months. Thus, we believe a device-side protection system has a significantly higher chance to be adopted.

Instead of implementing our protection system directly on the cellular connectivity hardware, we achieve protection by controlling the communication channel between the OS and the baseband. The smartphone is partitioned and the OS is separated from the baseband. The separation is implemented through virtualization. The actual core of our protection system is comprised of an AT command filter. With the implementation of our protection system based on the Android [34] platform we show that our approach is feasible for real-world smartphones. Still we think its design is general enough to be used for other smartphone OSes as well.

The main contributions of this chapter are:

- **Categorization of Signaling Issues:** We categorize different security and reliability issues that are caused by signaling traffic related to smartphone use and abuse. The issues can be separated into intentional actions (attacks), and side effects that can be abused for attacks.

- **Cellular Signaling Filter:** We introduce a novel mechanism to protect cellular network infrastructure against overloading from smartphones. This is achieved by filtering the signaling channel directly on the smartphone. This avoids expensive changes on the cellular core network. We further show that our novel security mechanism can be used to protect the user from Trojans that cause premium rate charges via SMS.

- **Safe-to-root virtualized Android:** We designed and built a safe-to-root virtualized Android. Our virtualized Android can be rooted and modified as the owner of the device wishes. The device manufacturer together with the operator retain full control over the cellular network interface (the baseband) and thus can prevent the device from being abused for launching attacks.

Chapter Organization

The rest of this Chapter is organized as follows. In Section 6.2 we give a detailed overview of the threats to both the network and the phone owner that are related to the baseband of a modern smartphone. The design of our protection system to mitigate these threats is described in Section 6.3. Implementation details of our prototype system are described in Section 6.4. In Section 6.5, we discuss our actual mitigation technique in great detail. The evaluation of our protection system is presented in Section 6.6. In Section 6.7, we discuss related work before we conclude and outline future improvements in Section 6.8.

6.2 Threats

In this section we introduce the different classes of threats that we address with the protection system presented in this work. There are three basic classes: threats that hijacked smartphones pose to the cellular core network, malware residing on the smartphone – with and without system privileges, and rooted devices.

6.2.1 Hijacked Phones and Mobile Botnets

The threats that hijacked phones and mobile botnets pose to the cellular network infrastructure and mobile customers is an emerging trend. A good example is the ikee.B [69] iPhone botnet. The bot infected about 22,000 devices and contained a HTTP-based Command and Control system.

Traynor et al. show in [85] that smartphone-based botnets can pose a serious threat to the cellular core network. They demonstrated that mobile botnets can overload backend systems such as the HLR and thus bring down the cellular network itself. Their attack is based on AT commands issued by zombie phones, which cause a high load on the HLR. Specifically, they issue the AT command to configure and enable call-forwarding settings. We discuss the actual details of the attack in Section 6.6 where we evaluate our protection system against the various attacks.

6 Improving the Security of the Cellular Modem Interface

The second issue with mobile botnets is their use of SMS messages for their Command and Control (C&C) communication, as we demonstrated in Chapter 5. Our Proof-of-Concept bot uses SMS messages for delivering the C&C messages between the nodes of the botnet. A similar Proof-of-Concept SMS controlled botnet was created in [90].

The SMS messages must be blocked to prevent botnet communication and to ensure that the subscriber (owner) does not incur any additional charges related to the increased SMS traffic.

6.2.2 PDP Context Change

Fast PDP context activation and de-activation leads to high network load on the GGSN and SGSN infrastructure. This is performed by either malicious applications or badly configured mobile phones. This is possible because on smartphone platforms such as Android any application has access to the network configuration and thus is able to change the packet-data settings.

On Android it is possible to force an PDP context change every 2 seconds. This will result in roughly 43,200 PDP activations per day (24 hours). A rogue application can easily carry out a Denial-of-Service attack against an operator's packet-data infrastructure, if it is installed on enough devices.

The GSM Association (GSMA) points out a similar problem [36] through the use of pre-paid SIM cards. Travelers who do not want to pay high roaming costs often buy pre-paid SIM cards. A flood of PDP context activation attempts can occur under two conditions: First, the pre-paid SIM card does not match the configured packet-data settings (the one of the home operator), but the phone keeps trying to activate packet-data every few seconds. Second, the pre-paid account is below the number of credits that are required to establish a PDP context. In both cases the PDP context creation is rejected by the network, but for the phone it looks like a technical error, and thus it repeatedly attempts to reconnect to the network.

6.2.3 Premium Rate SMS Trojans

Fraud caused by SMS Trojans such as `FakePlayer-A` [31] is a long standing problem in the mobile phone world costing consumers a considerable amount of money every year [65]. This kind of fraud is possible since on modern smartphones any application has access to the cellular API and is therefore able to send SMS messages. The same problem applies to voice calls to premium numbers.

Smartphone platforms such as Android or Symbian implement mandatory access control to restrict arbitrary access to system resources such as location, Internet, or cellular access. These permissions are hard coded into the application. At installation time of an application the user is shown a list of required permissions. The user can accept these or cancel the installation process. It is not possible to selectively accept or deny access privileges. Thus, many users simply accept such permission requests without considering their implications.

For example, on Android the permission required to send SMS messages is called *android.permission.SEND_SMS*.

6.2.4 Rooted Phones

Rooted or jailbroken smartphones are a serious security risk. Once a device is rooted, many security features of the operating system, such as network and cellular access restrictions as well as data-caging, are gone. Thus, the entry barrier for malware such as Trojans or botnets is much lower on rooted phones.

Rooting can happen in two ways. First, voluntarily by the owner who wants to be able to install additional, potentially unauthorized, applications. This type of rooting is often done by simply installing a modified firmware on the device. Thus, no security flaws are actually exploited.

Second, by malware such as DroidDream [53] in order to gain maximum privileges on the infected system. This type of rooting is achieved by exploiting known security flaws in the respective smartphone OS.

6.3 Design

Our aim is to mitigate Denial-of-Service attacks based on signaling traffic sent from mobile phones. As described in Section 2.4 the baseband is a phone's gateway to the cellular network. Consequently, our protection system must have exclusive control over the baseband hardware. For clarification, we define the following criteria for our protection system.

6 Improving the Security of the Cellular Modem Interface

Integrity Our protection infrastructure must withstand attacks from the smartphone OS. Even a rooted phone must not be able to directly tamper with the baseband. This can only be achieved, if our protection system is spatially *isolated* from the smartphone OS, e.g. it must not depend on its correct operation.

Completeness All cellular network access must be mediated and controlled by our trustworthy components.

Universality Our solution must be applicable to all cellular networks without requiring modifications to the operator's equipment.

Portability Our solution shall be usable on commercial off-the-shelf smartphones. It must not require additional hardware or hardware modifications. Our system has to support different baseband chips as well as popular smartphone CPUs. The solution must not depend on a certain smartphone OS. However, for practical reasons (open source, popularity) we chose the Android OS for this work.

Security Our protection system must not pose additional threats to the smartphone OS. This criterion is similar to the integrity criterion, but also includes availability and confidentiality of the whole system.

Upgrades and modifications to the cellular operator's equipment are very expensive and take a lot of time. In contrast, the smartphone market is advancing rapidly, with frequent releases of new smartphone generations. Each smartphone generation might bring new issues that would require new measures on the operator's side. We opted for a solution that addresses the signaling problem directly at its root, the smartphone itself (*Universality Criterion*).

A reasonable place for our protection system is the baseband as this is the smartphone's gateway to the cellular network (*Completeness Criterion*). The baseband processor has its own memory and is physically isolated from the application processor (*Integrity Criterion* and *Security Criterion*).

Baseband chipsets are under tight control by their manufacturers. Hardware details and the software stack are kept as trade secrets. Thus, no SDK or developer documentation is available. The *Portability Criterion* requires us to implement our protection

6.3 Design

system on commonly used basebands, which could turn out to be inherently difficult as the basebands might vary vastly. Also, a modified baseband would probably require re-certification, which due to time and cost constraints is infeasible.

Instead we chose to build our protection system on the application processor. The *Completeness Criterion* requires that the smartphone OS cannot directly access the baseband hardware. All cellular network access needs to be mediated by a custom proxy component. We call this component the *virtual modem*. The virtual modem runs as a separate task. We ensure spatial isolation between the smartphone OS, in our case Android, and the virtual modem by running Android in a virtual machine (VM). Figure 6.1 shows this setup. Direct hardware access of the Android VM to the baseband is denied. Instead we present the Android VM with an interface to the virtual modem. This ensures that even in the event of a rooted Android, the network operator cannot be adversely affected. For all other hardware, e.g. wireless LAN and graphics, we allow the VM exclusive access to the underlying hardware interfaces. We assume that the device's DMA feature can be restricted to safe memory locations[1]

6.3.1 Micro Kernel as Secure Foundation

In contrast to a monolithic kernel such as Linux a micro kernel merely implements essential mechanisms. This dramatically reduces the complexity of the kernel. Components such as device drivers or protocol stacks are implemented as user-level tasks [52]. Isolation between user-level tasks is enforced with address spaces. All communication between tasks is done via efficient explicit kernel-mediated inter-process communication (IPC).

Modern third-generation micro kernels implement object-capabilities. This access control scheme makes it possible to build systems that implement the principle of least authority (POLA). POLA states that each component is equipped with the minimum set of permissions necessary to fulfill its task.

The micro kernel ensures spatial and temporal isolation of its user-level tasks. It guarantees safe object access via object-capabilities. This ensures the *Integrity Criterion*.

---
[1] Technology such as IO-MMUs is already available in personal computers. Similar technology is likely to be implemented in future smartphone CPUs.

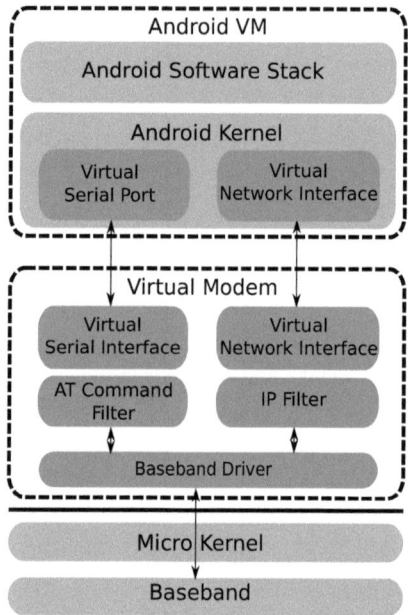

Figure 6.1: The architecture of our protection system. All interaction with the baseband is mediated by the *virtual modem*. Android runs inside its own virtual machine.

6.3.2 Virtualized Android

As outlined in the previous section the micro kernel partitions the system in a secure way. The partition running Android is implemented as a virtual machine.

Virtualization requires the Android OS to run with less privileges than the micro kernel. On the other hand, the Android kernel expects to have exclusive control of the hardware. Unfortunately, today's smartphone CPUs are not natively virtualizable, which prevents virtualization in the form of trap and emulate [68].

Fortunately, it is possible to run monolithic OS kernels such as Linux as a user-level task on top of a micro kernel. Härtig et al. [38] showed that the overhead of running a monolithic OS on top of a micro kernel is between 5 and 10 percent. We believe that this is acceptable on modern smartphones, given the merits it brings in terms of security.

In our setup the Android kernel is modified to run as an application on top of the micro kernel. As such, it can only access memory pages that we present it with. By granting a predefined set of IO memory pages, we can restrict the hardware that the Android kernel can access. We enforce that Android cannot directly access the baseband by not giving it access to the baseband's IO memory. Instead, we present it with an interface to our virtual modem. This ensures that all cellular communication is mediated by our protection system (*Completeness Criterion*).

Whereas we slightly modify the Android kernel, its user-level software stack remains unmodified. We designed the Android VM to be safe-to-root, be it voluntarily by the user, or by malware. If the user wants to flash his device, he is free to exchange the content of the Android partition. A commercial version of our protection system requires a bootloader that is capable of restricting updates to the Android partition. Required adjustments of currently used bootloaders are minimal.

6.3.3 Virtual Modem

The virtual modem is the only software that is allowed direct access to the baseband hardware. As such, it mediates all cellular network access of Android. It consists of the following components.

Baseband Driver The baseband driver contains all the logic needed to communicate with the baseband hardware. The actual implementation is specific to the device, and often contains numerous dependencies such as drivers for I2C or SPI buses. The baseband driver also contains the logic to tunnel IP data packets through the cellular data network.

Virtual Serial Interface Our virtual modem provides its client (the virtualized Android) with a virtual serial interface for sending and receiving the AT command stream.

AT Command Filter All AT commands are mediated and filtered by our AT command filter. The AT command filter is the central component that enforces our policies on the baseband. It will be explained in detail in Section 6.5.

Virtual Network Interface Once a data connection is established, all data packets are

6 Improving the Security of the Cellular Modem Interface

transfered between the baseband driver and Android via a virtual network interface.

IP Filter The virtual modem includes the infrastructure for network address translation (NAT).

6.4 Implementation

We built our prototype around an Intel x86-based smartphone. However, the design described in Section 6.3 applies equally well to the widely used ARM architecture.

We picked the *Fiasco.OC* [88] micro kernel as the foundation of our system. Fiasco.OC is a modern third-generation micro kernel, which provides the features outlined in Section 6.3.

6.4.1 Hardware

We developed our prototype on an Aava [8] development phone. The phone hardware is built around the Intel Moorestown [44] platform. It consists of a SoC that contains a graphics accelerator (GPU) and a low voltage Atom core. The Atom CPU is clocked at 1.5Ghz and supports hyperthreading. The board is equipped with 512MB RAM. For debugging purposes a UART is connected via SPI. The phone contains a ST-Ericsson U300 series baseband.

A picture of one of our development phones is show in Figure 6.2.

6.4.2 L4Android

The L4Android project [51] is based on L4Linux [87], a version of Linux that was ported from the machine interface to the micro kernel interface of Fiasco.OC. In addition to L4Linux, which is derived from the mainline Linux kernel, L4Android incorporates Google's Linux kernel modifications to support the Android software stack.

L4Android runs as an application in its own address space on top of the micro kernel. Each of the Android processes runs in its own address space and benefits of the same isolation capabilities as on the stock Android kernel. As the L4Android kernel ABI is compatible with Android, we can run all Android applications without modifications, even those containing native code.

6.4 Implementation

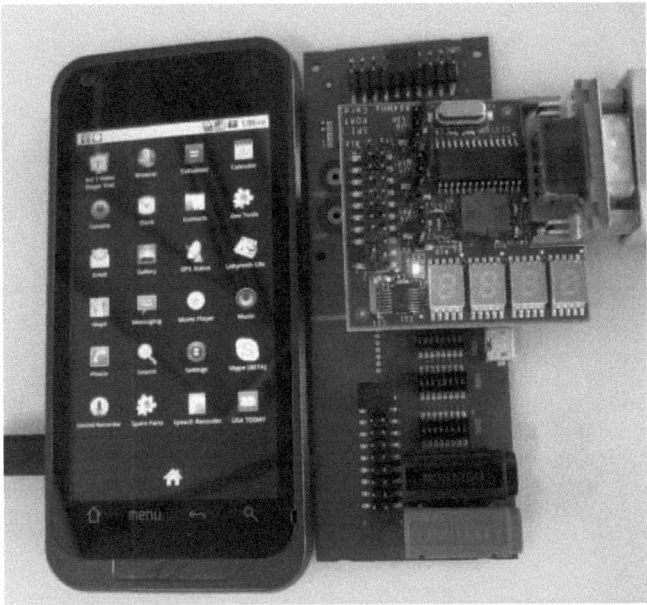

Figure 6.2: One of our Aava devices. A debug board is attached to the right and provides a serial line.

L4Android supports the Android user-level software stack in versions 2.1 (Eclair) up to 2.3 (Gingerbread) and enables us to even run multiple instances of Android in parallel on one device.

6.4.3 System Setup

Our setup is depicted in Figure 6.3. It is made up of two logical partitions: The *Android VM* and the *virtual modem* partition. The former runs the L4Android kernel and the Android user-level software stack (including applications).

The virtual modem partition consists of a L4Linux instance, the *Forwarder* and our *AT command filter*. We grant L4Linux exclusive access to the baseband. This has the benefit of allowing the use of the vendor supplied native Linux driver, and we do not need to implement our own one.

6 Improving the Security of the Cellular Modem Interface

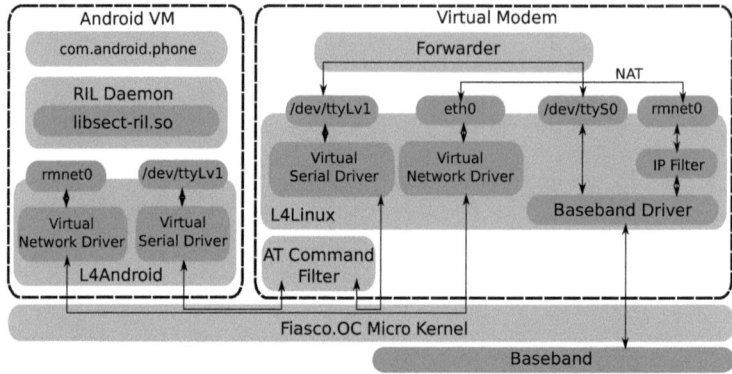

Figure 6.3: Our implementation consists of two components running on the Fiasco.OC micro kernel, the L4Android running the Android software stack and the virtual modem. The virtual modem is composed of three parts, the AT command filter, a Linux kernel that contains drivers and the Forwarder.

L4Linux is responsible for:

- Booting and initializing the baseband. This potentially includes loading a firmware to the baseband.

- Running the baseband driver. This includes the driver for the serial line that connects the baseband to the application CPU, and all the logic needed to demultiplex the serial stream into data packets and commands. Furthermore, it implements the protocol stacks needed to tunnel IP packets over the cellular network.

L4Linux implements advanced functionality such as IP filtering or Network Address Translation (NAT). It does not present a user interface as it does not require user interaction.

Communication between Android and the virtual modem partition is established via two channels. One is the virtual serial line to transmit the AT command stream. It is proxied by the AT command filter to implement the filtering. The other channel is the virtual network for IP-based data connections.

Virtual Serial Device The virtual serial device is used for all baseband commands. Both the L4Android and L4Linux kernel contain a custom driver that presents a serial device to applications. The custom driver establishes a virtual bidirectional serial line and sends all data via IPC.

Virtual Network Interface For data connections, we employ a shared memory based virtual Ethernet driver. Packets are written into a shared memory region, and the receiver is notified of incoming packets via IPC.

The L4Linux instance in the virtual modem partition forwards data received on the virtual devices to the corresponding physical device and vice versa. For the serial devices this task is performed by the Forwarder, which is implemented as a Linux application. It routes commands between the virtual serial line and the serial control channel of the baseband. In addition, the Forwarder parses the PDP context activation reply from the baseband (see Figure 6.4), extracts the parameters and applies them to the network interface presented by the baseband driver. The original values are replaced with the parameters necessary to configure the virtual network interface in the Android partition.

The forwarding of IP packets between the virtual network interface and the one provided by the baseband driver is performed using the Linux *netfilter* infrastructure. We setup a simple IP masquerading rule, but more advanced firewall rules can be added.

6.4.4 Modifications to the Android RIL

The Radio Interface Layer (RIL) daemon in Android abstracts details of the baseband implementation for upper layers in the Android stack. This includes voice calls, SMS messages, creation of PDP contexts, and configuration of the baseband. Specifics of the baseband are implemented by the baseband manufacturers in separate libraries, such as libreference-ril.so. The libraries are loaded by the RIL daemon to access the baseband functionality. Each vendor has to develop such a library when adopting Android to a new baseband.

From Android's perspective our virtual modem behaves like a specific baseband implementation. Consequently we built our own abstraction library (libsect-ril.so) for the RIL daemon. The rest of the Android user-level software stack remains unmodified.

The Android RIL configures the network interface used for data connections. As shown in Figure 6.4 the connection parameters of a PDP context are transfered as XML. Our library extracts these parameters and applies them to the virtual network interface.

6.5 The AT Command Filter

The baseband takes care of all interaction between the smartphone OS and the cellular network. The interface between the OS and the baseband is a serial character stream. The serial stream carries commands (signaling) and data (packet-data; IP packets). Voice is handled through other interfaces. Our focus is the signaling. Signaling is done via the GSM extensions to the AT command set as standardized in [5].

The curious reader might also think about the so-called GSM-codes that one can enter into a phone's dialer application (e.g. ##002# to clear call-forwarding settings). These GSM-codes are part of the Man-Machine Interface (MMI) standard [4] and are simply translated into AT commands by the user-level phone dialer application.

In the rest of this section we first characterize the signaling relevant AT commands and give some brief insights of our specific baseband. Then we discuss special issues with filtering AT commands and how we solved them. In the remaining part of the section we present our implementation, how we block commands, and how we profiled the AT commands to determine the baseline for configuring our filter.

6.5.1 AT Command Characterization

We analyzed the AT communication to determine what commands and command sequences are used to perform critical operations such as changing call-forwarding or packet-data settings. Below we briefly discuss the relevant commands.

AT+CGDCONT Configure a PDP context. This sets the connection parameters such as the Access Point Name (APN), user name and password, and other optional parameters. We provide an example of this command later in this section.

AT+CGACT Activate a configured PDP context. However, this standardized command, is not used on the ST-Ericsson baseband that our hardware comes with. There,

6.5 The AT Command Filter

activation of the PDP context works differently and is described below. There are other commands to activate and de-activate a PDP context, but these are not considered within the scope of this work.

AT*EPPSD PDP context control for our ST-Ericsson baseband. The command takes the PDP context index and the new state (1 = up or 0 = down) as arguments. In the next section we provide more details on the PDP context setup and activation.

AT+CMGS Send an SMS message. The SMS message is provided as hex encoded Protocol Data Unit (PDU). The command below sends an SMS message in PDU mode, the message consists of 17 bytes.

```
AT+CMGS=17
>
0001000c81101521436587000004d4f29c0e
```

ATD+4930835358585; Initiates a voice call to the given number. The semicolon signals the baseband that the call is actually a voice call. Without semicolon the baseband tries to establish a data call.

AT+CCFC Configure, activate, and de-activate call-forwarding settings. The command takes the type of call-forwarding such as when *busy* or *unreachable*, and the destination number as arguments. The example below sets call-forwarding for the *busy* state to the given number.

```
AT+CCFC=1,1,"4915112345678",129,0
```

AT+CFUN Configuration of the baseband state. The most common states are: Flight mode (stop all radio transmissions), GSM only, 3G only, and GSM+3G (prefer 3G) mode. The command below switches the baseband to Flight mode.

```
AT+CFUN=4
```

6.5.2 PDP Context Setup on the STE Baseband

First, the PDP context is configured using the standardized command `AT+CGDCONT`. Activation is performed by the custom `AT*EPPSD` command. The baseband replies with

6 Improving the Security of the Cellular Modem Interface

a XML text block containing the IP address, subnet mask, MTU, and DNS server IP addresses. Figure 6.4 shows an example of the whole process including the context configuration.

```
AT+CGDCONT=1,"ip","internet.t-mobile","",0,0
OK
AT*EPPSD=1,1,1
<?xml version="1.0"?>
<connection_parameters>
    <ip_address>10.165.132.86</ip_address>
    <subnet_mask>255.255.255.255</subnet_mask>
    <mtu>1500</mtu>
    <dns_server>193.189.244.225</dns_server>
    <dns_server>193.189.244.206</dns_server>
</connection_parameters>
OK
*EPSB
```

Figure 6.4: Configuration and activation of a PDP context on our ST-Ericsson baseband hardware

6.5.3 Special Problems

While analyzing the AT command interface and experimenting with our device we identified some additional issues with the AT commands.

Special case APN. Some operators have an additional APN for MMS, therefore, one has to take care of additional legal APN activate-deactivate sequences. Our implementation includes additional checks which ensure that deliverability of MMS messages is not restricted.

Command side effects. Certain AT commands have side effects that need to be taken into account by our filter. We determined that the baseband state switch command (AT+CFUN) is such a case. If the baseband is switched between 2G and 3G the PDP context is disconnected and reconnected.

6.5 The AT Command Filter

6.5.4 Filtering AT Commands

As shown in Figure 6.1 and 6.3, the AT command filter sits between the Android user-level telephony stack and the baseband.

The filter parses commands issued by the RIL (the RIL daemon runs in the Android partition) and enforces the configured filter rules. Commands that are not relevant are forwarded to the baseband without applying any parsing. Results are passed back to the RIL.

We implemented filters for all commands we discussed earlier in Section 6.5.1. These are: packet-data configuration and activation (AT+CGDMNT and AT*EPPSD), call-forwarding (AT+CCFC), modem control (AT+CFUN), SMS (AT+CMGS), and calls (ATD). The filter works as an intelligent rate limiter. It counts how often a command is issued within a period of time (the *interval*). If the count reaches the *threshold* all further commands issued within the *interval* are blocked. The rule below will allow issuing 5 AT+CCFC commands within 60 seconds.

```
AT_CCFC_interval = 60 (seconds)
AT_CCFC_threshold = 5 (# commands)
```

Certain commands have to be combined (see special issues Section 6.5.3). The core logic of our filter is shown in Figure 6.5.

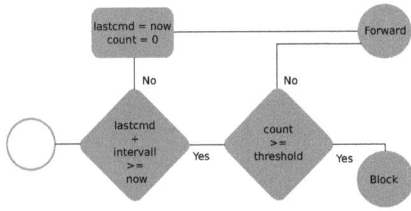

Figure 6.5: For the commands of interest we track each instance of a command within the configured interval. If the configured threshold is reached the command is blocked. When the interval expires the counter is reset.

101

6.5.5 SMS Filter

We implemented additional filters to inspect the PDU of SMS messages [6] sent from Android to the virtual modem, more closely. We implemented two features. First, a premium rate number detector. Second, a binary payload detector.

The **Short Code Detector** inspects the destination number of every SMS message that is sent. Premium rate numbers are mostly implemented using so-called *short codes*, telephone numbers as short as 4-6 digits. If a short code is detected and short code blocking is activated the command is blocked.

Our prototype blocks all SMS messages sent to short codes. To allow sending *legit* SMS to short codes we need to complete the implementation of the *secure GUI* we describe in the future improvements, in Section 6.8.1.

The **Binary Message Payload Detector** inspects the header and payload field of every SMS message that is sent through the filter. It uses a simple heuristic to determine if the message is binary. The heuristic checks one flag in the header and checks if the message body mainly consists of non-printable characters. The ratio of printable to non-printable can be configured. It further checks for base64 encoding and flags the message as binary if this is detected. If the message is determined to be binary it will be subjected to the rate limiting rule for binary SMS messages.

6.5.6 Blocking Commands

Commands are blocked by simply not forwarding them to the baseband. To not confuse the application logic in the RIL our filter issues an appropriate error message for each blocked command. The error message is injected into the stream that otherwise carries the responses from the baseband to the RIL. Special commands are never blocked due to various reasons. These are:

Switch to flight mode (AT+CFUN=4). This is necessary since flight mode is a required functionality that must always work. Even if the *threshold* for the CFUN command is reached a switch to flight mode is always permitted. In the worst case the phone will remain in flight mode until the *interval* expires.

PDP context deactivation (AT*EPPSD). This is necessary to prevent excessive data costs. For example, when the phone is roaming and the user wants to deactivate packet-data.

Emergency calls (ATD 911;) must always work due to regulations.

6.5.7 Profiling benign AT Command Usage

To determine useful and working intervals and thresholds for configuring our filter we monitored the AT commands we are interested in. To determine how often and when these commands are issued we set the intervals to 86,400 seconds (one day). Thus, the filter only counts the number of commands but never actually blocks anything. Table 6.1 shows AT command usage in general. Note, that the call-forwarding command (AT+CCFC) is issued multiple times at the point when the user opens the call-forwarding settings screen. This is because the phone always queries the network for the settings since these can be changed from multiple places. When the user changes a setting an additional command is issued. Followed by querying the state again. Therefore, the call-forwarding filter has to take into account that at a certain time multiple commands are executed in a row. Figure 6.6 shows an example output of our filter log right after the phone booted.

Command	#	When	Why
AT+CFUN	2	Boot	Flight mode. Normal mode.
AT+CFUN	1	Use	Switch to GSM-only.
AT+CDGMNT	1	Boot	Set PDP configuration.
AT*EPPSD	1	Boot	Activate PDP context.
AT+CMGS	1	Use	Send a SMS message.
ATD	1	Use	Issue a voice call.
AT+CCFC	3	Use	Query forwarding settings.
AT+CCFC	2	Use	Set a call-forwarding.

Table 6.1: AT commands issued during runtime.

6 Improving the Security of the Cellular Modem Interface

```
current time: 3793
APN[ 1]      : "internet.t-mobile"
     state : 1
     count : 1
     last  : 3793
     cfg   : 3793
APN current                          : 1
APN switch    count                  : 1
APN switch    interval (policy) : 86400
APN switch    threshold (policy): 6
APN switch    last                 : 3793
CALLFWD[0]    interval (policy) : 86400
CALLFWD[0]    threshold (policy): 5
CALLFWD[0]    last                 : 0
CALLFWD[0]    count                : 0
CALLFWD[1]    interval (policy) : 86400
CALLFWD[1]    threshold (policy): 5
CALLFWD[1]    last                 : 0
CALLFWD[1]    count                : 0
CALLFWD[2]    interval (policy) : 86400
CALLFWD[2]    threshold (policy): 5
CALLFWD[2]    last                 : 0
CALLFWD[2]    count                : 0
CALLFWD[3]    interval (policy) : 86400
CALLFWD[3]    threshold (policy): 5
CALLFWD[3]    last                 : 0
CALLFWD[3]    count                : 0
...           ...                    ...
CALLFWD[5]    count                : 0
GSMONLY       interval (policy) : 86400
GSMONLY       threshold (policy): 4
GSMONLY       last                 : 3778
GSMONLY       count                : 2
GSMONLY       mode                 : 1
BINSMS        interval (policy) : 86400
BINSMS        threshold (policy): 1
BINSMS        last                 : 0
BINSMS        count                : 0
SMSSHORT      count                : 0
```

Figure 6.6: The status of our AT command filter after booting the device for AT command profiling.

6.6 Evaluation

We developed a set of test applications to simulate rogue behavior such as updating call-forwarding settings or changing the PDP context. We further acquired a sample of an actual Premium Rate SMS Trojan for the Android platform to test against a real-world malware. Below we first describe our evaluation environment – our small GSM network, followed by the evaluation of our protection system against the threats we described in Section 6.2.

6.6.1 Our GSM Test Network

Our setup consists of a small GSM network that is based on an ip.access nanoBTS. The nanoBTS is managed by OpenBSC [92]. Our network is operated in a Faraday cage, where we conduct all our experiments safely. OpenBSC comes with additional components that provide a SGSN and a GGSN, which allows to operate a packet-data network in addition to the voice and SMS services. The setup allows us to monitor all relevant aspects of the cellular network. Such as PDP context establishment and incoming and outgoing SMS traffic.

Through the use of this environment we can test and verify our implementation.

6.6.2 Limiting the Call-forwarding Attack

The call-forwarding attack as described by [85] is based on insertion of call-forwarding settings by hijacked phones. Their attack requires 2,500 Transactions Per Seconds (TPS) for low traffic networks and up to 30,000 TPS for high traffic networks.

The victim phones issue AT+CCFC commands to configure and enable call-forwarding. The authors of [85] calculated that on average a command takes 4.7 seconds to complete, meaning one can issue up to 12 commands per minute. Thus, they require 11,750 bots to perform the attack on a low traffic network.

$$4.7\ seconds * 2,500\ TPS = 11,750\ hosts$$

For our initial experiment we configured the filter to allow 5 commands per minute. We chose this configuration because the Android call-forwarding configuration panel issues 3 commands when it is started. These commands query the network for the current state as shown in Table 6.1. The command that causes high load (enable call-forwarding) is only issued when the user changes a setting. After changing the setting

6 Improving the Security of the Cellular Modem Interface

the network is queried again. We, therefore, set the *threshold* = 5. With this setting the botnet's size already has to more than double in order to successfully perform the attack. Figure 6.7 shows the necessary size increase of the botnet described in [85] to perform the attack if the zombie phones are equipped with our protection system.

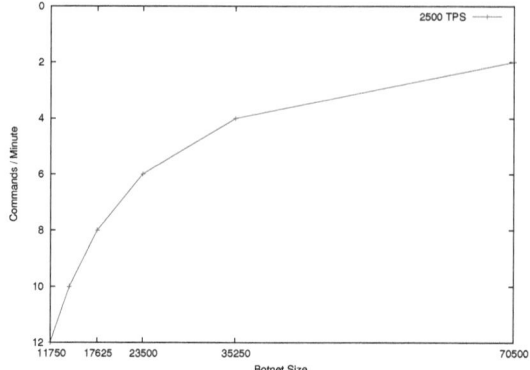

Figure 6.7: The increase in size of the botnet necessary to maintain the 2,500 TPS with our protection system in place.

To further improve the protection provided by our system we can increase the *interval* of the call-forwarding filter, resulting in an even lower number of commands per minute. For example, allowing just 10 call-forwarding commands over a period of 10 minutes of time. Such a threshold results in 1 command per minute on average, which is reasonable for normal usage.

6.6.3 Limiting PDP Context Changes

To limit the number of PDP context changes we have to mediate two different commands. The commands are described in Section 6.5.3 where the side effects of these commands are presented. The side effect, which must be detected by our system, is switching the baseband mode between GSM-only, 3G-only, and GSM+3G. Calculating the *threshold* for PDP context changes is straightforward.

Defining p_t as the threshold for PDP context changes, e_t as the threshold for AT*EPPSD

6.6 Evaluation

commands, and c_t as the threshold for AT+CFUN commands, yields

$$p_t = e_t + c_t.$$

The graph in Figure 6.8 shows the number of possible PDP context changes depending on the settings of p_t. Without any rate limiting applied, 30 changes per minute is the maximum possible.

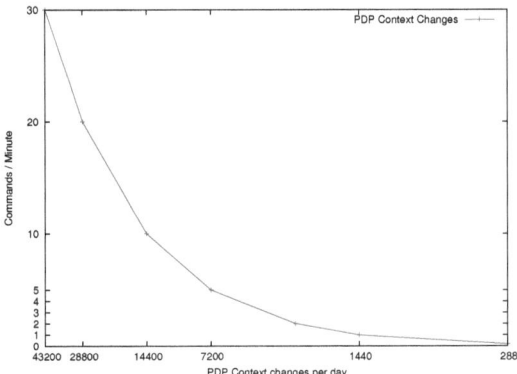

Figure 6.8: The graph shows the number of PDP context changes possible based on how many commands are allowed per minute. The last point at 288 PDP context per day is based on 0.2 commands per minute.

6.6.4 SMS Trojan

We installed the `FakePlayer-A` [31] premium SMS Trojan to a test phone equipped with our protection system. The Trojan is built to send a SMS message to a premium rate number to steal money from the victim. We closely monitor the log output of our filter to determine what happens. In addition we also monitor the phone's behavior on our private GSM network, to see if it actually sends any SMS message or not.

The Trojan tries to send an SMS message to the number 3353. As this number is short (only 4 digits) it is detected by our *Short Code Detector*. Extensive analysis of this Trojan [31] determined that the Trojan sends a SMS to either 3353 or 798657.

6 Improving the Security of the Cellular Modem Interface

However, we only observed attempts to 3353 in our lab. Below is an excerpt from our filter log.

```
AT+CMGS=15
00010004813335000006b71cce56bb01

number: 3353
short phone number >3353< detected, could be premium
filterd: filtered returned: 0
filterd: blocking >00010004813335000006b71cce56bb01<
```

6.6.5 SMS Controlled Botnets

SMS controlled botnets such as discussed in Chapter 5 send and receive SMS messages for their command and control channel. Since this work focuses on outgoing signaling traffic we decided to only look at outgoing traffic, from the phone to the network.

To prevent botnet communication we enabled our *Binary Payload Detector* together with the rate limiting for the AT+CMGS command (the command to send SMS messages). The rate limiter will only prevent the phone from sending binary SMS messages at a high rate. Binary SMS messages are rarely sent by the user since these are mostly used by applications. Furthermore, the most common usage of binary SMS are messages that are received by the phone (e.g. as part of MMS). Text messages on the other side are often used in an instant messaging scheme with a high rate of outgoing and incoming messages. Therefore, blocking text messages will be more complicated since they would need to be analyzed thoroughly before one is able to safely block them.

6.7 Related Work

Related work falls into four categories. First, security enhancements for smartphones. Second, virtualization on smartphones. Third, Android specific security extensions. Forth, infrastructure-based security enhancements for cellular networks.

Traynor et al. summarize in [83] the lack of security features on mobile and smartphones and discuss possible solutions. Part of their work presents SELinux [54] as means of access control of system resources. But the authors come to the conclusion that such an approach is infeasible. In our work, we directly address the specific problems of signaling attacks. We not only propose a solution, but fully implemented a prototype and evaluated it. Mulliner et al. [64] build a label based tracking system that tracks a process' access to network interfaces to limit future access to other network resources such as the cellular modem. The SEIP [94] architecture uses D-Bus in combination with SELinux to enforce access policies for applications accessing various system resources on a smartphone.

Selhorst et al. [75] describe a Trusted Mobile Desktop prototype that, similar to our approach, uses a micro kernel together with multiple virtualized Linux instances. In their setup, a so-called *User Linux* partition drives the baseband and runs the user's applications. A separate component signs and encrypts SMS. The encrypted SMS is then sent via the User Linux's baseband driver. They do not provide means for protecting the cellular network from malicious behavior of the User Linux partition. Schmidt et al. [74] describe how a trusted mobile platform can be built on a trustworthy platform. The authors employ the Turaya security kernel to run a virtualized legacy operating system (Linux) side-by-side with multiple trusted engines. They do not propose to control network interfaces. Klein et al. [50] design and implement seL4. In their work the authors demonstrate that the implementation of a modern third-generation micro kernel can be formally verified to match its specification. VMware [67] ported their virtualization software to the Android platform. However, this port runs a virtualized guest version of Android on top of a host Android. Additionally, our solution is based on a micro kernel and has a significantly smaller trusted computing base.

Enck et al. build TaintDroid [25] a taint tracking based security and privacy enhancement for Android. TaintDroid is able to track which data an application accessed. The MockDroid [14] Android enhancement adds the possibility to selectively mock specific features such as Internet connectivity. Thus applications cannot use specific functionalities even if they actually are available.

Previous work on protecting cellular phone networks has targeted other attack vectors such as [84] that investigates countermeasures for preventing resource exhaustion attacks against cellular phone networks carried out over the Internet.

6.8 Conclusions

We designed and implemented our protection system called the *virtual modem* to protect cellular network infrastructure from hijacked smartphones. The virtual modem mediates all signaling traffic from the smartphone OS to the baseband and thus protects

6 Improving the Security of the Cellular Modem Interface

the cellular network. The implementation is based on running Android and our virtual modem in isolated partitions on top of a micro kernel. Our solution is independent from the baseband and thus supports wide adoption.

We evaluated our implementation using real mobile phone hardware that we connect to our own GSM network. Our GSM network allows us to monitor all relevant activities of the phone. Part of the evaluation was installing a real-world Trojan on the device. The Trojan was successfully launched, but our virtual modem prevented the fraudulent access to the cellular network.

Signaling attacks are a serious threat and recognized as such by the GSMA. The evaluation of our protection system showed that it can effectively prevent these attacks and thus protect cellular core networks. In addition it protects the end-user.

6.8.1 Future Improvements

Our virtual modem can be enhanced with the following functionality.

VPN Gateway The modem can establish access to VPNs in a way that even a rooted Android cannot access the key material.

Advanced Intrusion Detection/Prevention We can enhance our IP filter with logic to detect and prevent attacks against the smartphone as well as against the operator.

Policy Update Infrastructure The virtual modem can include an update infrastructure to allow the operator to update the filtering policies. Such an update would be performed transparently to the user.

Secure GUI With the addition of a *secure graphical user interface*, we can implement a dialog that enables the virtual modem to ask the user for admission of premium SMS and calls. To make sure that Android malware cannot mimic the admission dialog, or automatically send the confirmation input, the dialog must be presented in a way that does not depend on Android for input. Doing so requires virtualization of the graphics and input hardware.

Hardware Virtualization Porting the Android kernel to our micro kernel requires a significant amount of work. When hardware support for CPU virtualization becomes available on smartphones, it can both reduce the amount of modifications to the Android kernel, and may improve the performance of our Android VM.

6 Improving the Security of the Cellular Modem Interface

7 Conclusions

In this work, we argued that the interface of the cellular modem is one key element in the puzzle for securing mobile phones and cellular communications. So far the cellular modem interface has not gotten any attention in terms of security. This work has described several approaches for analysis, attacks, and defense of the cellular modem interface of cellular handsets and especially smartphones.

Throughout this work, we analyzed the interface between the modem and the mobile operating system in order to utilize it to conduct vulnerability analysis of SMS implementations on smartphones. Through our analysis, we discovered several security issues in the analyzed targets. We further investigated feature phones where the cellular modem and the application processor (that runs the mobile operating system) are executed on the same CPU. Here, we determined that software bugs on the application processor affect the modem processor in a negative way. This results in security and reliability issues that can be abused for Denial-of-Service attacks. In the next step of our work, we investigated the possibilities of malware that has unlimited access to the cellular modem of a smartphone. We implemented and evaluated a Proof-of-Concept cellular botnet that communicates through SMS messaging. The investigation showed that such a botnet is feasible. Our investigation on defending against this kind of cellular botnet showed that mobile operators will have to play a primary role in this. Following up on our botnet investigation we conceived a protection mechanism that mitigates attacks carried out by malware that has unlimited access to the cellular modem. Our protection system mediates the communication channel between the mobile operating system and the cellular modem. The primary goal is to prevent Denial-of-Service attacks against cellular networks carried out from hijacked smartphones. Furthermore, it is able to prevent fraud and SMS-based botnet communication.

In this work we proved that the modem interface plays an essential role in securing mobile phones and cellular networks. The interface can be abused for vulnerability analysis and likewise for malicious activities, as presented in Chapters 3, 4, and 5. This shows the importance of the cellular modem interface and confirms our claim *(i)*. Our second claim *(ii)* states that access and usage of the cellular modem is only controlled in a very coarse-grain way. We validated this claim in Chapters 5 and 6 where we discuss malicious usage of the modem. We show that once an application has access to the modem it can freely interact with the modem. Malware that has access to the cellular modem can leverage all its capabilities and abuse it for carrying out attacks. Based on the previous observation we validated claim *(iii)* where we state that unlimited access poses a security threat to the handset and the cellular network. Unlimited access can be abused for botnet communication, premium rate SMS fraud, or Denial-of-Service attacks. Based on the validation of our three claims we proved that our main hypothesis is valid – that the overall security of cellular handsets and cellular networks can be strengthened through improving the security of the cellular modem interface. In Chapter 6 we presented an implementation of a strong access and usage control mechanism that improves the overall security of both, the handset and the network.

We believe that protecting the cellular modem interface from unlimited access from the mobile operating system is the necessary next step in the evolution of securing mobile handsets. Our work will help leading towards creating more responsible devices that participate in the growing world of cellular communication.

Future Work

We plan to enhance our virtual modem by adding features like intrusion detection and prevention. We believe this is a promising path for future research. Likewise, we want to further investigate mobile malware. Specifically mobile malware that contains botnet-like functionalities such as remote control are becoming a serious threat to mobile phone users and mobile network operators. Malicious code on smartphones is an ongoing trend that has to be followed.

Further directions will concentrate on the cellular modem itself, since only very little work has been carried out on this topic so far. This includes vulnerability analysis of the software running on the modem CPU itself and incorporate protocol level security measures directly into the modem software.

We want to extend our work beyond mobile handset and investigate other areas of cellular devices. Especially the area of cellular Machine-to-Machine (M2M) communication is very interesting.

7 Conclusions

Acknowledgements

I am extremely grateful to my advisor Jean-Pierre Seifert who provided me with an excellent environment for perusing my research. He always supported my independent research and let me select my own research topics. His motivation and support during my work was always greatly appreciated. This work would not have been possible without his support.

Next I want to thank my various research collaborators. Especially Nico Golde and Charlie Miller for their work on the SMS projects; without them the projects would not have been such great success. Furthermore, I want to thank Steffen Liebergeld, Matthias Lange, and Dmitry Nedospasov for their collaboration and support in general.

I want to thank my colleagues and friends from the *Security in Telecommunications* (SECT) group for making this journey fun and productive. Special thanks go to Patrick Stewin for many helpful discussions during the years of my research.

Furthermore, I want to thank my former advisor, colleagues and friends from the SecLab at the University of California at Santa Barbara where I started working on (smartphone) security while working on my Master's degree many years ago. This work still benefits from things I learned from my time there.

Additional thanks go to the many other people who helped and supported me in various ways during my time as a PhD student and while writing this work. This includes the time before I joined TU Berlin. The incomplete list is (in no particular order): Volker Roth, Bernhard Ager, Sebastian Schinzel, and Michael Kasper.

Very special thanks go to my parents. Thanks for your support and everything else!

Acknowledgements

Finally, I want to thank Deutsche Telekom AG and specifically the people at the Laboratories not only for general funding, but also for providing an outstanding work environment.

List of Figures

2.1	The basic setup of a cellular network.	15
2.2	The basic design of a modern smartphone.	17
3.1	SMS_DELIVER Message Format	22
3.2	The User Data Header (UDH)	23
3.3	Unsolicited AT result code that indicates the reception of an SMS message	24
3.4	Logical model of our injector framework	26
3.5	The UDH for SMS Concatenation	29
3.6	The UDH for SMS Port Addressing	29
4.1	Our setup: A laptop that runs OpenBSC and the fuzzing tools, the nanoBTS, and some of the phones we analyzed.	41
4.2	Mobile terminated SMS	44
4.3	Logical view of our setup	46
4.4	Format of the SMS_SUBMIT PDU	47
4.5	The User Data Header	48
4.6	Timing of SMS message delivery attempts	58
5.1	Kademlia P2P C&C	72
5.2	SMS only C&C	74
5.3	SMS-HTTP hybrid C&C	76
5.4	Basic command structure	80
6.1	The architecture of our protection system. All interaction with the baseband is mediated by the *virtual modem*. Android runs inside its own virtual machine.	92
6.2	One of our Aava devices. A debug board is attached to the right and provides a serial line.	95

List of Figures

6.3 Our implementation consists of two components running on the Fiasco.OC micro kernel, the L4Android running the Android software stack and the virtual modem. The virtual modem is composed of three parts, the AT command filter, a Linux kernel that contains drivers and the Forwarder. 96

6.4 Configuration and activation of a PDP context on our ST-Ericsson baseband hardware 100

6.5 For the commands of interest we track each instance of a command within the configured interval. If the configured threshold is reached the command is blocked. When the interval expires the counter is reset. 101

6.6 The status of our AT command filter after booting the device for AT command profiling. 104

6.7 The increase in size of the botnet necessary to maintain the 2,500 TPS with our protection system in place. 106

6.8 The graph shows the number of PDP context changes possible based on how many commands are allowed per minute. The last point at 288 PDP context per day is based on 0.2 commands per minute. 107

List of Tables

4.1	Mobile phone manufacturer market shares	39
5.1	Connectivity times.	69
6.1	AT commands issued during runtime.	103

List of Tables

Bibliography

[1] 3GPP/ETSI. 3GPP TS 03.38 Alphabets and language-specific information. `http://www.3gpp.org/ftp/Specs/html-info/0338.htm`, 1998.

[2] 3GPP/ETSI. 3GPP TS 03.40 Technical realization of the Short Message Service. `http://www.3gpp.org/ftp/specs/html-info/0340.htm`, 1998.

[3] 3GPP/ETSI. 3GPP TS 04.11 Point-to-Point (PP) Short Message Service (SMS) Support on Mobile Radio Interface. `http://www.3gpp.org/ftp/specs/html-info/0411.htm`, 1998.

[4] 3GPP/ETSI. 3GPP TS 02.30 - Man-Machine Interface (MMI) of the Mobile Station (MS). `http://www.3gpp.org/ftp/Specs/html-info/0230.htm`, June 2002.

[5] 3GPP/ETSI. 3GPP TS 27.007 - Technical Specification Group Terminals; AT command set for User Equipment (UE). `http://www.3gpp.org/ftp/Specs/html-info/27007.htm`, June 2003.

[6] 3GPP/ETSI. 3GPP TS 23.040 - Technical realization of the Short Message Service (SMS). `http://www.3gpp.org/ftp/Specs/html-info/23040.htm`, September 2004.

[7] 3GPP/ETSI. 3GPP TR 23.840 Study into routeing of MT-SMs via the HPLMN. `http://www.3gpp.org/ftp/Specs/html-info/23840.htm`, 2007.

[8] Aava Mobile Oy. Aava Mobile. `http://www.aavamobile.com/`.

[9] ABI Research. Worldwide Mobile Subscriptions Number More than Five Billion. `http://www.abiresearch.com/press/3531-Worldwide+Mobile+Subscriptions+Number+More+than+Five+Billion`, October 2010.

Bibliography

[10] T. Ahonen. Mobile Phone Market Shares for year of 2009. http://communities-dominate.blogs.com/brands/2010/02/phone-market-shares-for-year-of-2009-and-last-quarter-2009.html, February 2010.

[11] T. Ahonen. *Tomi Ahonen Almanac 2010 Mobile Telecoms Industry Review*. February 2010.

[12] P. Amini and A. Portnoy. Sulley - Pure Python fully automated and unattended fuzzing framework. http://code.google.com/p/sulley/.

[13] BBC News. Estonia hit by 'Moscow cyber war'. http://news.bbc.co.uk/2/hi/europe/6665145.stm, 2007.

[14] A. R. Beresford, A. Rice, N. Skehin, and R. Sohan. MockDroid: trading privacy for application functionality on smartphones. In *12th Workshop on Mobile Computing Systems and Applications*, March 2011.

[15] S. Byers, A. D. Rubin, and D. Kormann. Defending against an Internet-based attack on the physical world. *ACM Trans. Internet Technol.*, 4(3):239–254, 2004.

[16] cellular-news. A "rising Tide" of SS7 Based Mobile Network Fraud. http://www.cellular-news.com/story/46377.php, November 2010.

[17] CERT. Advisory CA-1998-01 Smurf IP Denial-of-Service Attacks. http://www.cert.org/advisories/CA-1998-01.html, January 1998.

[18] Clickatell (Pty) Ltd. Clickatell Bulk SMS Gateway. http://www.clickatell.com.

[19] ComScore. German Mobile Market Share. http://www.comscore.com/index.php/Press_Events/Press_Releases/2010/1/comScore_Reports_November_2009_German_Mobile_Market_Share, November 2010.

[20] ComScore. U.S. Mobile Subscriber Market Share. http://comscore.com/Press_Events/Press_Releases/2010/7/comScore_Reports_May_2010_U.S._Mobile_Subscriber_Market_Share, May 2010.

[21] D. Danchev. DDoS extortion-themed scam circulating. http://www.zdnet.com/blog/security/ddos-extortion-themed-scam-circulating/7180, August 2010.

Bibliography

[22] J. de Haas. Mobile Security: SMS and a little WAP. http://www.itsx.com/hal2001/hal2001-itsx.ppt, August 2001.

[23] Deutsche Telekom Medien GmbH. Das Örtliche. http://dasoertliche.de.

[24] J. R. Douceur. The Sybil Attack. In *Revised Papers from the First International Workshop on Peer-to-Peer Systems*, IPTPS '01, 2002.

[25] W. Enck, P. Gilbert, B.-G. Chun, L. P. Cox, J. Jung, P. McDaniel, and A. N. Sheth. TaintDroid: an information-flow tracking system for realtime privacy monitoring on smartphones. In *Proceedings of the 9th USENIX conference on Operating systems design and implementation*, OSDI'10, pages 1–6, Berkeley, CA, USA, 2010. USENIX Association.

[26] W. Enck, P. Traynor, P. McDaniel, and T. La Porta. Exploiting Open Functionality in SMS-Capable Cellular Networks. In *Conference on Computer and Communications Security*, 2005.

[27] T. Engel. Remote SMS/MMS Denial of Service - Curse Of Silence. http://berlin.ccc.de/~tobias/cursesms.txt, December 2008.

[28] European Telecommunications Standards Institute (ETSI). GSM 06.06 (ETS 300 642): Digital cellular telecommunication system (Phase 2); AT Command set for GSM Mobile Equipment (ME). http://www.etsi.org, 1999.

[29] R. Farrow. DNS Root Servers: Protecting the Internet. *Network Magazin*, 2003.

[30] Federal Communications Commission. Cognitive Radio Technologies and Software Defined Radios (ET Docket No. 03-108; FCC 07-66), June 2007.

[31] Fortinet. Trojan-SMS.AndroidOS.FakePlayer-A. http://www.fortiguard.com/encyclopedia/virus/android_fakeplayer.a!tr.html, August 2010.

[32] F. Freiling, T. Holz, , and G. Wicherski. Botnet Tracking: Exploring a Root-Cause Methodology to Prevent Distributed Denial-of-Service Attacks. In *Proceedings of 10th European Symposium On Research In Computer Security (ESORICS'05)*, July 2005.

Bibliography

[33] GigaOM. When It Comes to Apps, Feature Phones Are the New Black. `http://gigaom.com/2010/03/27/when-it-comes-to-apps-feature-phones-are-the-new-black/`, May 2010.

[34] Google Inc. Android. `http://www.android.com/`.

[35] J. B. Grizzard, V. Sharma, C. Nunnery, B. B. H. Kang, and D. Dagon. Peer-to-peer botnets: Overview and case study. In *Proceedings of the Workshop on Hot Topics in Understanding Botnets*, April 2007.

[36] GSM Association (GSMA). Network Efficiency Threats v0.4a, May 2010.

[37] C. Guo, H. J. Wang, and W. Zhu. Smartphone attacks and defenses. In *Third ACM Workshop on Hot Topics on Networks*, 2004.

[38] H. Härtig, M. Hohmuth, J. Liedtke, J. Wolter, and S. Schönberg. The performance of µ-kernel-based systems. In *Proceedings of the sixteenth ACM symposium on Operating systems principles*, SOSP '97, pages 66–77, New York, NY, USA, 1997. ACM.

[39] Hay Systems Ltd. HSL Mobile Messaging Gateway. `http://www.hslsms.com`.

[40] W. J. Hengeveld. Windows Mobile AT-command log-driver. `http://nah6.com/~itsme/cvs-xdadevtools/itsutils/leds/logdev.cpp`.

[41] T. Holz, M. Steiner, F. Dahl, E. Biersack, and F. Freiling. Measurements and Mitigation of Peer-to-Peer-based Botnets: A Case Study on Storm Worm. In *Proceedings of the First USENIX Workshop on Large-Scale Exploits and Emergent Threats (LEET '08)*, 2008.

[42] R. Hund, M. Hamann, and T. Holz. Towards Next-Generation Botnets. In *4th European Conference on Computer Network Defense (EC2ND 08)*, 2008.

[43] IDC. Western European Mobile Phone Market Grows. `http://www.idc.com/getdoc.jsp?containerId=prUK22402810`, June 2010.

[44] Intel Corporation. Introducing the Next-Generation Intel® Atom™ Processor-based Platform. `http://download.intel.com/pressroom/kits/atom/z6xx/pdf/Fact_Sheet_Intel_Atom_Processor_Platform.pdf`, 2010.

Bibliography

[45] V. Iozzo and R.-P. Weinmann. iPhone Safari vulnerability allowed to steal the SMS database. http://news.cnet.com/8301-27080_3-20001126-245.html, March 2010.

[46] ip.access Ltd. nanoBTS 1800. http://www.ipaccess.com/picocells/nanoBTS_picocells.php.

[47] D. Johnson, A. Menezes, , and S. A. Vanstone. The elliptic curve digital signature algorithm (ecdsa). *Int. J. Inf. Sec.*, 1(1):36–63, 2001.

[48] B. Jurry. Siemens Mobile SMS Exceptional Character Vulnerability. http://www.xfocus.org/advisories/200201/2.html, January 2002.

[49] M. Kenney. Ping of Death. http://insecure.org/sploits/ping-o-death.html, October 1996.

[50] G. Klein, K. Elphinstone, G. Heiser, J. Andronick, D. Cock, P. Derrin, D. Elkaduwe, K. Engelhardt, R. Kolanski, M. Norrish, T. Sewell, H. Tuch, and S. Winwood. seL4: Formal Verification of an OS Kernel. In *ACM Symposium on Operating System Principles*, pages 207–220. ACM, 2009.

[51] M. Lange and S. Liebergeld. L4Android: Android on top of L4. http://www.l4android.org, 2011.

[52] J. Liedtke. On micro-kernel construction. In *Proceedings of the fifteenth ACM symposium on Operating systems principles*, SOSP '95, pages 237–250, New York, NY, USA, 1995. ACM.

[53] Lookout Inc. DroidDream. http://blog.mylookout.com/2011/03/security-alert-malware-found-in-official-android-market-droiddream/, March 2011.

[54] P. Loscocco and S. Smalley. Integrating Exible Support For Security Policies Into The Linux Operating System. In *Proceedings of the FREENIX Track of the USENIX Annual Technical Conference*, 2001.

[55] P. Maymounkov and D. Mazières. Kademlia: A peer-to-peer information system based on the xor metric. In *Revised Papers from the First International Workshop on Peer-to-Peer Systems*, IPTPS '01, pages 53–65, London, UK, 2002. Springer-Verlag.

Bibliography

[56] Micromax. Micromax mobile. `http://www.micromaxinfo.com`.

[57] C. Miller. Exploiting the iPhone. `http://securityevaluators.com/content/case-studies/iphone/`, August 2007.

[58] C. Miller, M. Daniel, and J. Honoroff. Exploiting Android. `http://securityevaluators.com/content/case-studies/android/index.jsp`, October 2008.

[59] C. Miller and C. Mulliner. Fuzzing the Phone in your Phone. `http://www.blackhat.com/presentations/bh-usa-09/MILLER/BHUSA09-Miller-FuzzingPhone-SLIDES.pdf`, August 2009.

[60] Mobile Security Lab. SonyEricsson WAP Push Denial of Service. `http://www.mseclab.com/?page_id=123`, January 2009.

[61] D. Moore, V. Paxson, S. Savage, C. Shannon, S. Staniford, and N. Weaver. Inside the Slammer Worm. *IEEE Security and Privacy*, 1:33–39, 2003.

[62] B. Müller. From 0 to 0-Day On Symbian. `https://www.sec-consult.com/files/SEC_Consult_Vulnerability_Lab_Pwning_Symbian_V1.03_PUBLIC.pdf`, 2009.

[63] C. Mulliner and G. Vigna. Vulnerability Analysis of MMS User Agents. In *Proceedings of the Annual Computer Security Applications Conference (ACSAC)*, Miami, FL, December 2006.

[64] C. Mulliner, G. Vigna, D. Dagon, and W. Lee. Using Labeling to Prevent Cross-Service Attacks Against Smart Phones. In *Proceedings of the Conference on Detection of Intrusions and Malware, and Vulnerability Assessment (DIMVA)*, volume 4064 of *LNCS*, pages 91–108, Berlin, Germany, July 2006. Springer.

[65] J. Niemelä. Mobile Malware And Monetizing 2011. `https://noppa.tkk.fi/noppa/kurssi/t-110.6220/luennot/T-110_6220_mobile_maleware.pdf`, 2011.

[66] Oulu University Secure Programming Group. PROTOS Security Testing of Protocol Implementations. `http://www.ee.oulu.fi/research/ouspg/protos/`, 2002.

[67] PC World. VMWare Shows off Mobile Virtualization on Android. http://www.pcworld.com/article/219671/vmware_shows_off_mobile_virtualization_on_android.html, 2011.

[68] G. J. Popek and R. P. Goldberg. Formal requirements for virtualizable third generation architectures. *Commun. ACM*, 17:412–421, July 1974.

[69] P. A. Porras, H. Saidi, and V. Yegneswaran. An Analysis of the iKee.B iPhone Botnet. In *Proceedings of the 2nd International ICST Conference on Security and Privacy on Mobile Information and Communications Systems (Mobisec)*, May 2010.

[70] R. Racic, D. Ma, and H. Chen. Exploiting MMS vulnerabilities to stealthily exhaust mobile phone's battery. In *Proceedings of the Second IEEE Communications Society / CreateNet International Conference on Security and Privacy in Communication Network (SecureComm)*, Baltimore, MD, Auguest 28 - September 1, 2006.

[71] Routo Telecommunications Ltd. Routo Messaging. http://www.routomessaging.com.

[72] Samsung. S3C6400. http://www.samsung.com/global/system/business/semiconductor/product/2007/8/21/661267ptb_s3c6400_rev15.pdf, 2007.

[73] J. H. Schiller. *Mobile Communications (second edition)*. 2003.

[74] A. U. Schmidt, N. Kuntze, and M. Kasper. On the deployment of mobile trusted modules. In *Wireless Communications and Networking Conference, 2008. WCNC 2008. IEEE*, pages 3169–3174. IEEE, 2008.

[75] M. Selhorst, C. Stüble, F. Feldmann, and U. Gnaida. Towards a trusted mobile desktop. In *Proceedings of the 3rd international conference on Trust and trustworthy computing*, TRUST'10, pages 78–94, Berlin, Heidelberg, 2010. Springer-Verlag.

[76] K. Singh, S. Sangal, N. Jain, P. Traynor, and W. Lee. Evaluating Bluetooth as a Medium for Botnet Command and Control. In *Proceedings of the International*

Bibliography

Conference on Detection of Intrusions and Malware, and Vulnerability Assessment (DIMVA), July 2010.

[77] Strategy Analytics. Q3 2009 Cellular Baseband Market Review. http://blogs.strategyanalytics.com/HCT/post/2010/03/09/Q3-2009-Cellular-Baseband-Market-Review.aspx, March 2010.

[78] SUN Microsystems. Java Micro Edition. http://www.oracle.com/technetwork/java/javame/index.html.

[79] Texas Instruments. OMAP3 Processors: OMAP3430. http://focus.ti.com/general/docs/wtbu/wtbuproductcontent.tsp?templateId=6123&navigationId=12643&contentId=14649.

[80] The Honeynet Project. Honeynet Project. http://project.honeynet.org, 2005.

[81] The Intrepidus Group. WebOS: Examples of SMS delivered injection flaws. http://intrepidusgroup.com/insight/2010/04/webos-examples-of-sms-delivered-injection-flaws/, April 2010.

[82] TopNews.in. Micromax becomes the third largest handset manufacturer in India. http://www.topnews.in/micromax-becomes-third-largest-handset-manufacturer-india-2260105, April 2010.

[83] P. Traynor, C. Amrutkar, V. Rao, T. Jaeger, P. McDaniel, and T. La Porta. From Mobile Phones to Responsible Devices. *Journal of Security and Communication Networks (SCN)*, 2010.

[84] P. Traynor, W. Enck, P. McDaniel, and T. La Porta. Mitigating Attacks on Open Functionality in SMS-Capable Cellular Networks. *IEEE/ACM Transactions on Networking (TON)*, 2009.

[85] P. Traynor, M. Lin, M. Ongtang, V. Rao, T. Jaeger, T. La Porta, and P. McDaniel. On Cellular Botnets: Measuring the Impact of Malicious Devices on a Cellular Network Core. In *ACM Conference on Computer and Communications Security (CCS)*, November 2009.

[86] P. Traynor, P. Mcdaniel, and T. La Porta. On attack causality in internet-connected cellular networks. In *Proceedings of the USENIX Security Symposium*, 2007.

Bibliography

[87] TU Dresden. L4Linux - Running Linux on top of L4. http://os.inf.tu-dresden.de/L4/LinuxOnL4/, January 2011.

[88] TU Dresden. The Fiasco microkernel. http://os.inf.tu-dresden.de/fiasco/, January 2011.

[89] WAP Forum. WAP-209-WSP Wireless Application Protocol MMS Encapsulation Protocol. http://www.wapforum.com, 2002.

[90] G. Weidman. Transparent Botnet Control for Smartphones over SMS. http://www.grmn00bs.com/GeorgiaW_Smartphone_Bots_SLIDES_Shmoocon2011.pdf, January 2011.

[91] R.-P. Weinmann. All Your Baseband Are Belong To Us. https://cryptolux.org/media/hack.lu-aybbabtu.pdf, 2010.

[92] H. Welte. OpenBSC. http://openbsc.osmocom.org/trac/, 2008.

[93] O. Whitehouse. Nokia Phones Vulnerable to DoS Attacks. http://www.infoworld.com/article/03/02/26/HNnokiados_1.html, February 2003.

[94] X. Zhang, J.-P. Seifert, and O. Acicmez. SEIP: Simple and Efficient Integrity Protection for Open Mobile Platforms. In *Information and Communications Security*, volume 6476 of *Lecture Notes in Computer Science*, pages 107–125. Springer Berlin / Heidelberg, 2010.

i want morebooks!

Buy your books fast and straightforward online - at one of world's fastest growing online book stores! Environmentally sound due to Print-on-Demand technologies.

Buy your books online at
www.get-morebooks.com

Kaufen Sie Ihre Bücher schnell und unkompliziert online – auf einer der am schnellsten wachsenden Buchhandelsplattformen weltweit! Dank Print-On-Demand umwelt- und ressourcenschonend produziert.

Bücher schneller online kaufen
www.morebooks.de

VDM Verlagsservicegesellschaft mbH
Heinrich-Böcking-Str. 6-8   Telefon: +49 681 3720 174   info@vdm-vsg.de
D - 66121 Saarbrücken   Telefax: +49 681 3720 1749   www.vdm-vsg.de

Printed by Books on Demand GmbH, Norderstedt / Germany